D0113727

Biofuels Refining
and Performance

ABOUT THE EDITOR

Ahindra Nag, Ph.D., is a Senior Assistant Professor in the Department of Chemistry at the Indian Institute of Technology, Kharagpur. He has 21 years of teaching experience and has published 60 research papers in major national and international journals. He is the author of three other books: *Analytical Techniques in Agriculture, Biotechnology, and Environmental Engineering; Environmental Education and Solid Waste Management;* and *Foundry Natural Product Materials and Pollution.*

Biofuels Refining and Performance

Ahindra Nag, Ph.D.   Editor

New York   Chicago   San Francisco   Lisbon   London   Madrid
Mexico City   Milan   New Delhi   San Juan   Seoul
Singapore   Sydney   Toronto

The **McGraw·Hill** Companies

Library of Congress Cataloging-in-Publication Data

Biofuels refining and performance / Ahindra Nag, editor.
    p.   cm.
   Includes bibliographical references and index.
   ISBN 978-0-07-148970-6 (alk. paper)
    1. Biomass energy.   2. Biotechnology.   3. Renewable energy sources.
  4. Energy development.   I. Nag, Ahindra.
  TP339.B5437   2008
  662′.88—dc22   2007039373

Copyright © 2008 by The McGraw-Hill Companies, Inc. All rights reserved. Printed in the United States of America. Except as permitted under the United States Copyright Act of 1976, no part of this publication may be reproduced or distributed in any form or by any means, or stored in a data base or retrieval system, without the prior written permission of the publisher.

1 2 3 4 5 6 7 8 9 0   FGR/FGR   0 1 3 2 1 0 9 8 7

ISBN 978-0-07-148970-6
MHID 0-07-148970-3

Printed and bound by Quebecor/Fairfield.

This book is printed on acid-free paper.

McGraw-Hill books are available at special quantity discounts to use as premiums and sales promotions, or for use in corporate training programs. For more information, please write to the Director of Special Sales, McGraw-Hill Professional, Two Penn Plaza, New York, NY 10121-2298. Or contact your local bookstore.

Sponsoring Editor
  Larry S. Hager

Production Supervisor
  Richard C. Ruzycka

Editing Supervisor
  Stephen M. Smith

Project Manager
  Rasika Mathur

Copy Editor
  Yumnam Ojen

Proofreader
  Anju Panthari

Indexer
  WordCo Indexing Services

Art Director, Cover
  Jeff Weeks

Composition
  International Typesetting and
  Composition

Information contained in this work has been obtained by The McGraw-Hill Companies, Inc. ("McGraw-Hill") from sources believed to be reliable. However, neither McGraw-Hill nor its authors guarantee the accuracy or completeness of any information published herein, and neither McGraw-Hill nor its authors shall be responsible for any errors, omissions, or damages arising out of use of this information. This work is published with the understanding that McGraw-Hill and its authors are supplying information but are not attempting to render engineering or other professional services. If such services are required, the assistance of an appropriate professional should be sought.

Contents

Contributors

Sebastian Bojanowski *Department of Math, Natural Science, and Information Technology, Laboratory for Waste Treatment Processes, University of Applied Sciences Giessen-Friedberg, Giessen, Germany* (CHAP. 8)

K. B. De *Department of Chemistry, Indian Institute of Technology, Kharagpur, India* (CHAP. 1)

M. P. Dorado *Department of Physical Chemistry and Applied Thermodynamics, EPS, University of Cordoba, Cordoba, Spain* (CHAP. 4)

B. B. Ghosh *Department of Mechanical Engineering, Indian Institute of Technology, Kharagpur, India* (CHAP. 7)

Keikhosro Karimi *Department of Chemical Engineering, University of Technology, Isfahan, Iran* (CHAP. 3)

Gerhard Knothe *National Center for Agricultural Utilization Research, Agricultural Research Service, U.S. Department of Agriculture, Peoria, Illinois* (CHAP. 5)

P. Manchikanti *Agriculture Engineering Department, Indian Institute of Technology, Kharagpur, India* (CHAP. 2)

Ahindra Nag *Department of Chemistry, Indian Institute of Technology, Kharagpur, India* (CHAPS. 2, 6, 7)

A. K. Sinha *Department of Electrical Engineering, Indian Institute of Technology, Kharagpur, India* (CHAP. 9)

Ernst A. Stadlbauer *Department of Math, Natural Science, and Information Technology, Laboratory for Waste Treatment Processes, University of Applied Sciences Giessen-Friedberg, Giessen, Germany* (CHAP. 8)

Mohammad J. Taherzadeh *School of Engineering, University of Borås, Borås, Sweden* (CHAP. 3)

Preface

The continuous use of the world's crude oil reserve and a corresponding escalation in its price together with the limited coal reserves have stimulated the hunt for renewable sources of energy. The main sources of renewable energy are biomass, biogas, methanol, ethanol, and biodiesel; solar active (photovoltaic), solar passive (preheating of water), wind, mini hydel, and mini tidal are important sources which produce less pollution and protect the environment.

Much attention has been given to biomass and its modifications as a substitute for fossil fuels in the Western world. Among the modifications are biogas, alcohol, biodiesel, and manure. Presently, electrical power is attractive in many respects and the search is on for renewable and nonfinite resources to produce and supplement electrical energy.

The first chapter discusses energy and its biological sources. If biofuel is one of the expected solutions, we must know where is the beginning of the crisis and its solution. This chapter reviews the background story along with an optimistic outlook for a safe energy resource on our green earth. The second chapter discusses energy from photosynthetic plants and their inherent recycling nature, as well as the environmental benefits involved. These sources of energy are the solution for energy management. The third chapter discusses bioethanol, which is now one of the main actors in the fuel market. Its market grew from less than a billion liters in 1975 to more than 39 billion liters in 2006, and is expected to reach 100 billion liters in 2015. The chapter discusses the variety of raw materials, such as sugars, starch, and lignocellulosic substances, that produces bioethanol and also covers some of the market issues. To extend the use of biodiesel, the main concern is the economic viability of producing biodiesel. Edible oils are too valuable for human feeding to run automobiles. So, the emphasis must be on low-cost oils, i.e., nonedible oils, animal fats, and used frying oils. There are many nonedible feedstock crops growing in underdeveloped and developing countries; biodiesel programs here would give multiple social and economic benefits.

The fourth chapter discusses different plant sources used for production of biodiesel, properties of biodiesel, and processing of vegetable oils as biodiesel, and compares engine performance with different biodiesels.

Biodiesel is the methyl or other alkyl esters of vegetable oils, animal fats, or used cooking oils. Biodiesel also contains minor components such as free fatty acids and acylglycerols. Important fuel properties of biodiesel that are determined by the nature of its major and minor components include ignition quality and exhaust emissions, cold flow, oxidative stability, viscosity, and lubricity. The fifth chapter discusses how the major and minor components of biodiesel influence the mentioned properties.

Different techniques of biodiesel preparation and resulting engine performance are discussed in detail in Chap. 6. The seventh chapter discusses ethanol and methanol as fuel in the internal combustion engine and emphasizes their advantages (such as a higher octane number) over gasoline. Cracking of lipids turns polar esters into nonpolar hydrocarbons. This is accompanied by a fundamental change in physical and chemical properties. Products formed give rise to new applications in the fuel sector and for chemical commodities, e.g., detergents. The eighth chapter explores routes to provide these alternative hydrocarbons from lipids. It concentrates on substrates (seeds, vegetable oils, animal fat) and conversion pathways as well as analytical tools.

The ninth chapter discusses the fuel cell, an electrochemical device and nonpolluting alternative energy source that converts the chemical energy of a fuel (hydrogen, natural gas, methanol, gasoline, etc.) and an oxidant (air or oxygen) into electricity with water and heat as by-products.

The book is organized in a manner to cater to the needs of students, researchers, managerial organizations, and readers at large. We welcome the reader's opinions, suggestions, and added information, which will improve future editions and help readers in the future. Readers' benefits will be the best reward for the authors.

AHINDRA NAG, PH.D.

Biofuels Refining
and Performance

Energy and Its Biological Resources

K. B. De

1.1 Energy (Yesterday, Today, and Tomorrow)

Today's energy concept needs to be modified and should be presented as an integrated management-oriented approach. For example, the problem of nutrition of human population and livestock is also an important item in the energy inventory. So the per capita energy requirement will include 2000 kcal of the basal requirement in the form of nutrients; amount of energy required to produce that amount of food; energy required to preserve the food; energy required to collect the daily requirement of 600–800 L of water; energy required for washing, cleaning, and bathing; and energy required for lights, fans, air conditioners, and transport.

Today's energy concept should also include the awareness that heat is a wasteful form of energy, always downhill, and hence efficiency is at the most 30–40% and that of an automobile is as low as 15–20%. Even if we go modern, a solar photovoltaic panel has an efficiency of 8%, a solar thermal power plant has 15%, and from sunshine to electricity through biomass is only 1%.

In order to establish innovative technologies for highly effective utilization of solar light energy, fundamental research is being conducted in the following areas:

1. **Dye-sensitized solar cells:** New types of dye-sensitized solar cells mimicking the active sites of the natural photosynthesis system.

2. **Artificial photosynthesis:** Hydrogen production from water, using metal oxide semiconductor photocatalyst systems and effective fixation of CO_2 by metal or metal complex catalysts.

Estimated contribution of renewable energy resources in the United States by AD 2000, excluding hydro- and geothermal energy, amounts to approximately 5% of the estimated total consumption of 100 quads. Tropical countries, namely, India, receive 1648–2108 kWh/m^2 of solar energy in different parts with 250–300 days of sunshine, most of which is unutilized.

While shifting our attention from today into the future, we should look at some discussions that took place in the 12th Congress of World Energy, a conference held in New Delhi, during September 18–23, 1983, the main themes of which were management, policy, development, and quality of life. There were four divisions, and each of these divisions had four sections containing 157 technical papers. In the concluding session, Dr. J. S. Foster, chairman of the program committee, on behalf of the International Executive Council, gave a summary.

1. **Innovation:** Commenting on innovation, a report from Israel narrated absorption refrigeration, and Austria reported on a thermal power plant, investigating a treble Rankine cycle using three separate working fluid loops. Brazil reported methane from urban refuge and collaborative international efforts on controlled nuclear fusion were highlighted.

2. **Self-Reliance:** Self-reliance has been well emphasized.

3. **Diversification:** Diversification in national or regional supply ensures a robust energy structure, reducing vulnerability to vagaries of nature, resource, or market fluctuations.

4. **Dependence:** Dependence on fossil fuels can be reduced with proper substitution by biogas, solar, wind, and nuclear powers.

5. **Efficiency and conservation:** Waste heat recovery, cogeneration, and recycling of energy were in the technological aspect. Public and social consciousness through education is the other aspect.

6. **Development:** International cooperation and development assistance should involve mainly (a) financial resources, (b) technology transfers, and (c) transfer of managerial and engineering skills.

7. **Care of the environment:** Pollutions from fossil fuels, nuclear reactors, and effects on forests and vegetation from dams are to be studied along with future expansion schemes.

8. **Quality of life:** Indiscriminate and unplanned use of energy may lead to negative and harmful impacts. Need for energy education

and man power training with an integrated approach has been recommended.

9. **Urgency:**
   a. World population will reach 10 billion in 2020.
   b. Half of the population will have only 20 GJ/yr.
   c. The other half in the industrialized countries will use 15 times as much, i.e., 300 GJ/yr.
   d. "Firewood crisis" has changed to "firewood catastrophe" and the forest cover is diminishing globally at the rate of 250,000 km^2/yr. Energy administration in developing countries depends mainly on three denominators:
      (1) Growth rate of population
      (2) Energy self-reliant populations growing in size but lowest rate
      (3) Rural population largest in size, but lowest rate of energy consumption in most countries

Recommendations of new and alternative energy resources are available. The emphasis is on nonfossil and renewable resources, namely, biogas and biomass, solar active (photovoltaic), solar passive (preheating of water), wind, minihydroelectric, and minitidal resources. The major attempts for conservation include: conservation side legislation, education, awareness, management, and forecasting.

In Europe and South America, biomass and its modification have been given a lot of attention as a substitute to fossil fuels. The primary material, of course, is the waste of different plant and vegetable origins. The conversions are to biogas, alcohol, and manure. Proper selection of waste material may lead to optimal production of the right transform. Advanced countries, point out that electrical power is attractive in many respects and that the search for renewable and infinite resources to produce and supplement electrical energy should continue. Hydropower, solar energy, wind, solid waste, biomass, geothermal energy, ocean tidal power, and ocean thermal gradients are a few resources that need attention. In fact, many institutions and organizations have created demonstration models for these.

In the United Kingdom, the emphasis seems to be more on proper selection of local conditions and availability of the resources. Biomass and biogas need collection, transport, and processing to be properly useful for energy generation. Setting up aerogenerators, wind pumps, and solar heating will depend on available and favorable conditions and the proper location. If successfully implemented, they can reduce the local demand or share the load of a national power grid. The other resources that remain to be developed and commercialized are listed in the following discussion.

1. **Fusion of thermonuclear devices** (an application of plasma physics):

$$^2_1D + {}^2_1D \rightarrow {}^3_2He + n + 3.2\,\text{Mev} \qquad {}^2_1D + {}^3_1T \rightarrow {}^4_2He + n + 17.6\,\text{Mev}$$

$$^2_1D + {}^2_1D \rightarrow {}^3_1T + {}^1_1H + 4.0\,\text{Mev} \qquad {}^6_3Li + n \rightarrow {}^4_2He + {}^3_1T + 4.8\,\text{Mev}$$

$$\overline{{}^2_1D + {}^3_1T \rightarrow {}^4_2He + n + 17.6\,\text{Mev} \qquad {}^2_1D + {}^6_3Li \rightarrow 2{}^4_2He + 22.4\,\text{Mev}}$$

$$^2_1D + {}^3_2He^3 \rightarrow {}^4_2He^4 + {}^1_1H + 18.3\,\text{Mev}$$

$$\overline{6{}^2_1D \rightarrow 2{}^4_2He + 2{}^1_1H + 2n + 43.1\,\text{Mev}}$$

Several designs and modifications are suggested:

$$2P \rightarrow e^+ + \gamma + D \qquad D + P \rightarrow T + \gamma \qquad 2T \rightarrow He + 2P$$

The fusion reaction, omnipresent in the sun, needs to be tried out:

$$2H^1 \rightarrow e^+ + \nu + H^2$$

where two protons fuse, and deuterium, positron, and neutrino are evolved; energy is evolved in two steps; four protons are annihilated for each helium formed. Much of the reaction mechanism is yet unknown, but the model shows great promise.

2. **Geothermal source:** Other than volcanic or geyser origin at an 8000-ft depth of the earth's crust, it is possible to obtain geothermal steam at 2000°C, which can be used for producing electricity.

   Hot dry rock (HDR) remains out of reach at present capability of drilling. But "heat mining," as estimated by Los Alamos Scientific Laboratory, promises 1.2 cents/MJ compared to 2 cents/MJ from an oil-fired thermal plant ($34/bbl).

3. **Aerodynamic generations:** Several models are available. Low-velocity windmills are also being used. Wind is stronger at upper atmosphere; array of floating windmills are also designed.

4. **Hydrodynamics:** High hopes are created by some hydroelectric firms, who proclaim that power can be effectively generated by ocean waves and ocean currents.

5. **Magnetohydrodynamic generators:** High-temperature combustion gas expands through a nozzle where ionized sodium is introduced and directed to a magnetic field and a moving conductor cuts the field, and an electromagnetic field (EMF) is produced.

6. **Oil shale and oil sand:** Though of limited supply, these have not been fully explored.

7. **Coal conversion**: Many models for fluidization and gasification of coal are available.

8. **Black box or hydrogen fuel cell**: Usually, these use hydrogen as input fuel based on reverse hydrolysis (see last part of Sec. 1.6):

   At anode: $H_2 \rightarrow 2H^+ + 2e^-$

   At cathode: $O_2 + 4H^+ + 4e^- \rightarrow 2H_2O$

9. **Hydrogen as fuel**: Hydrogen as fuel is gaining popularity. The most common sources are from (a) excess of nuclear energy, (b) windmills, (c) hydroelectric power, (d) biological sources to some extent, (e) fuel cell (see Sec. 1.6), and (f) microbial hydrogen production (see Sec. 1.16).

10. **Biological energy**: A number of biological energy transformation principles, very attractive, remain at the conceptual state.

1.2 Energy

A body can do work, or work can be done upon a body; a body of water can turn a turbine, or one may pedal a bike to move it. If work is done on a body, it will possess energy. When energy is possessed by a body, the body can do work.

An agent may do work when it possesses energy, i.e., the amount of work that an agent can do is the amount of energy it possesses. So a body may gain kinetic and potential energy or lose the gained energy by producing heat or converting it to other forms of work.

Kinetic energy is due to the motion of a body.

Potential energy is due to the position or status of a body.

Frictional or colligative motion energy is produced in a waterfall; heat evolves to overcome a frictional resistance or checks the motion of a body but sets useless motion to others (e.g., rolling of pebbles in a stream or dust behind a vehicle). Mechanical friction causes a matchstick to ignite.

Units of energy are the same as those of work and are assigned equivalent quantities. Some important definitions and units are given in the appendix. Energy content of some common substances are provided in Table 1.1.

1.2.1 Thermodynamics

All three principles of thermodynamics are very much applicable in the area of biological energy and chemical changes related to it. It is worthwhile to review a few fundamental points. Chemical reaction can take

TABLE 1.1   Energy Content of Some Common Substances

Food value or fuel value	
Food value	Fuel value, kcal/g
Carbohydrates	4
Proteins	5
Fats (lipids)	9
Plant biomass (wet)	2
Plant biomass (ash free, dry)	4.5
Animal biomass (wet)	2.5
Animal biomass (ash free, dry)	5.5
Coal	7.0 [3200 kcal/lb]
Gasoline	11.5 [42,000 kcal/gal]

Average need for an adult human as consumer			
	Personal or survival need	Total social and establishment need	Ratio (total:personal)
Air	300 cuft/d	5000 cft/d	17:1
Water	0.66 gal/d	2000 gal/d	3030:1
Water (nonreturn)	0.3 gal/d	750 gal/d	2500:1
Energy (food)	1×10^6 kcal/yr	87×10^6 kcal/yr	87:1
Land (vegetarian food)	0.3 acre/yr	0.6 acre/yr	
Land (nonvegetarian food)	0.3 acre/yr	4 acre/yr	
Plant body other than food			

1 ton dry weight per year = 1-acre forest (and/or 3 tall trees of 12-in. diameter or 15 small trees of 6-in. diameter)

place only if the energy status changes, i.e., A will be converted to B only if B has a free energy content less than that of a change in free energy ΔF that is easy and spontaneous; reactions may be written as

$$A = B + (-\Delta F) \quad \text{or} \quad A = B - \Delta F$$

or

$$-\Delta F = F_A - F_B$$

The reaction is called exergonic, or energy is evolved or given out. If ΔF has a positive expression, the reaction is driven by the input of energy and called endergonic; such reactions are difficult to complete. At equilibrium, $\Delta F = 0$ ($\pm$), a point which may be arrived at by the end of the reaction, or a reaction may be typically of that type (practically sluggish, the progress of the reaction will depend on the change in concentration of reactants, the change of temperature or pressure, etc.).

$\Delta F = \Delta F_0 + RT \ln B/A$, where B/A is the ratio at equilibrium or equilibrium constant, i.e., K_{eq}. Then, $0 = \Delta F_0 + RT \ln B/A$ or $\Delta F_0 = - RT \ln B/A =$

$-1363 \log_{10} K_{eq}$ at 25°C. Here, $R = 1.987$ cal/mol/K, $T = (273 + 25)$ K = 298 K, and $\ln B/A = 2.303 \log_{10} K_{eq}$. This expression can be very useful:

K_{eq}	$\log_{10} K_{eq}$	$\Delta F_0 = -1363 \log_{10} K_{eq}$
1×10^0	0	0
$1 \times 10^{\pm 1}$	± 1	± 1363
$1 \times 10^{\pm 2}$	± 2	± 2726
$1 \times 10^{\pm 3}$	± 3	± 4086

When A and B exist equimolar, then the expression $\Delta F = \Delta F_0 + RT \ln 1$ means $\Delta F = \Delta F_0$, and the state is called a *standard state*.

Chemical conversions and change of state need some other consideration in the light of the third law of chemical thermodynamics:

$$\Delta F = \Delta H - \Delta TS$$

ΔH is the change in heat content, T is the absolute temperature at which the reaction occurs, and ΔS is the change in entropy (change, GR), or degree of disorder in the system, understood as the heat gained isothermally and reversibly per unit rise of temperature at which it happens (unit being calories per kelvin). The absolute value of H and S of a system cannot be directly determined. "Heat content" is also known as "heat content at constant pressure" or "enthalpy." The third law suggests chemical pathway of finding entropy values in absolute terms. The first law of thermodynamics deals with conservation of energy and the second law with the relation between heat and work.

1. Energy cannot be destroyed or created, i.e., the sum of all energies in an isolated system remains constant.
2. All systems tend to approach a state of equilibrium. This means that the entropy change of a system depends only on the initial and final stages of the system, expressed by R. Clausius.
   a. The total amount of energy in nature is constant.
   b. The total amount of entropy in nature is increasing.

1.3 Energy-Dependent Ecosystems

All forms of life are dependent on availability of energy at all levels, the creation, growth, and maintenance (defense, offense, and survival). The requirement and utilization of energy are mainly in two forms; the most important are nutrient and environmental energy in the form of heat and light.

It is easy to observe that extremely cold or hot regions are not favorable for the growth of living things. Likewise, the absence of light limits the propagation and proliferation of photosynthetic biotic species.

The sun, of course, radiates energy into space of which only an insignificant part is shared by this planet of ours called Earth. Because of its spin and its orbital rotation, a seasonal variation occurs in the total insolation on the earth's surface, which averages approximately 20 kcal/(m^2 · yr). The incident radiation comprises 2000–8000 Å, 50% of which is in the visible range (3700–7700 Å); only a small part of the incident energy is utilized by living systems.

Solar constants are given as 1.968 cal/(cm^2 · min) = 3.86 × 10^{33} erg/s = 1.373 kW/m^2. There are variations in the figures, depending on the source of information. However, the energy received on the earth's surface is mostly thermal and wasted. Biological fixation is restricted to photophosphorylation.

Let us look at the components of ecosystems that are capable of utilizing incident energy and some interrelationships between them.

Autotrophs (meaning self-surviving), also known as producers, mainly the photosynthetic systems, are the largest users of sunlight. Theoretically, anywhere there is light they should grow, provided other inputs are favorable. In arid land, the lack of nutrients; in deserts, the lack of water; and at higher-altitude, low temperatures, low CO_2 tension and other adverse conditions will prevent the proliferation of autotrophs, leaving otherwise sufficient insolation unutilized (energy fixation by photosynthetic pathway is treated elsewhere). Producers growing on detritus (dead organic materials) are not well described in the literature, but these could be autotrophs.

Heterotrophs (mixed surviving or unlike surviving), on the other hand, survive partly depending on the nutrient sources made available by other living systems. Most animals are heterotrophic. Therefore, animals are also called consumers.

If animals survive mainly on autotrophic materials, they are called primary consumers, commonly known as herbivores. If animals largely survive on other animals as their source of food, they are called secondary consumers, popularly known as carnivores. Predators are animals that hunt their animate food, known as prey. The prey–predator relationship plays an important role in nature and contributes to the ecologic balance.

1.3.1 Photosynthetic factors

Assuming that the wavelength of light remains constant, the intensity influences the rate of photosynthesis, which is why the earlier part of the forenoon is the most productive, and higher intensity of light energy and higher temperature slow down the photosynthetic rate. Likewise, a cloudy day does not slow down the normal photosynthetic rate of particular species to any observable extent.

Metabolically speaking, reports are insufficient to conclude anything based on this observation, even though the above information itself is very useful and valuable. At the onset of daybreak, the photosynthetic machinery gets into action after a dark rest period and the rate is at its peak; the carbon dioxide tension (partial pressure) at the immediate microenvironment is also higher (it is yet to be established that higher carbon dioxide tension facilitates photosynthesis, though the reverse is true). As the reaction proceeds with time, all other conditions remaining the same, the anabolic machineries including the enzymes and coenzymes (particularly NADP/Co II system) are fully occupied and ATP systems are also fully utilized. ATP production is, in turn, dependent on respiration (oxidative process), which to some extent is competitive with carbon fixation. Geological and geographical factors contribute greatly to ATP productivity.

Let us turn again to the consideration of biogeological and biogeographical distribution on energy. For an energy-based ecosystem, the biosphere may be classified into two major types: terrestrial, and aquatic. These can also be subdivided into eight intraterrestrial types: terrestrial, subterrestrial, epilimnon, mesolimnon, hypolimnon, estuarine, epimarine, and submarine. What do these have to do with our objective? Natural distribution of flora and fauna largely depend upon the types of microenvironments mentioned above.

At this point, it need not be assumed that the arctic belt, being very cold, is biologically unproductive. The author was surprised to see the existence of almost a minitropical pocket, 66° north latitude and 20° east longitude (Jockmock, Sweden) due to uninterrupted insolation for almost 90 days and prolonged daylight for 60 more days. The flora and fauna have adapted to survival techniques for the cruelty of adverse nature during the long, dark winter months.

1.4   Bioenergy

Energy can be derived from living systems in restricted forms only. Lignocelluloses are burned to get heat, and vegetable oils are often used for illumination. These may also serve as nutrients for different biotic species in various forms, i.e., cellulose, starch, and sugars. In other words, chemically stored energy may be reused in the form of fuel (firewood) or nutrients (food, feed, fodder, etc.). Animals can be employed to do different mechanical work. Animals directly (fish, meat) or indirectly (egg, milk) may provide nutrients for others. Use of dried cow dung as cooking fuel in rural areas is also a well-known example of animal products indirectly contributing to this field. But examples of direct energy flow from living systems are still in the conceptual state. Scientists dream that, one day, light emitted by fireflies or high voltage generated by electric eels may be of great use in the near future.

The production of alcohol or methane by microbial fermentation of common plant wastes are well-known phenomena. Recently however, scientists have started looking into these phenomena with greater interest, so that, in either gas or liquid form, their production and use can be optimized and made efficient. Plant bodies have been used as antennas, and plant leaves have been demonstrated to work as batteries. The survival of all biotic species depends directly or indirectly on solar energy. Studying the energy-based ecosystem raises awareness of this fact. Obviously, the most common question becomes: If the sun happens to be the source of all energy, why then is the solar energy not harnessed by different devices? There are inherent limitations of most of the physical devices by (a) way of efficiency, (b) critical cost, (c) maintenance, (d) reliability, and (e) other factors.

In photosynthetic systems operating in green vegetations of the above points, (b), (c), and (d) are enormously better. Its characteristic limitations are [for point (a)] the incident insolation, the ability to use only a narrow spectrum [for point (e)], and requiring the proportionate amount of soil surface area for insolation, optimal nutrients, temperature, and moisture in the microenvironment. Here nature provides several mutants from which we can take, pick, screen, or select the most tolerant variety. We may resort to genetic engineering for tissue cultures or selective hybridization.

What is our objective? Along with the effort to harness the solar energy by different physical methods, parallel efforts of optimal use of solar energy through biotic fixation should be attempted. This involves understanding the following:

1. The living world in its entirety, i.e., ecology.

2. The photosynthetic systems in different species: terrestrial, aquatic, or mixed.

3. Application of the above to develop science and technology for:
   a. Better management of the biotic systems useful for our purpose
   b. Conversion of biological raw materials into energy rich products

4. Coordination for quality of life, pollution abatement, and sparing of nonrenewable resources for future generations.

   A few examples that may not be out of place include potato, tomato, eucalyptus, and so forth. Though of wild origin, they have been appreciated and have been cultivated for this use after studying and admiring their productivity and receptivity. Later by scientific manipulation, new strains have been developed for cultivation.

It is justified to discuss certain established facts for making sufficient conceptual clarity for special topics. Some aspects of energy relations in living systems will be discussed in detail. Some other aspects will not be discussed in detail because existing "know-how" is rather limited.

1.5 Biological Energetics

The study of bioenergetics leads us into a world of novelty and greater significance and has found new encouragement in industry. The biogas generation by anaerobic fermentation has also led to new interest in research in the light of bioenergetics.

The study of energy relations for each chemical step in the living system may be an item of bioenergetics. The energy change can be calculated in terms of calories or joules per mole. This is applicable for catabolic processes, for example, the anaerobic or glycolytic paths or oxidative phosphorylation. The anabolic paths are equally fitting, e.g., the carbon fixation or the photosynthesis and nitrogen fixation by the symbiotic organisms [1].

The accounting and balancing of free energy change of certain reactions may lead to some fruitful conclusions. When glucose is oxidized in a bomb calorimeter (an almost one-step reaction),

$$C_6H_{12}O_6 + 6O_2 \rightarrow 6CO_2 + 6H_2O - 686,000 \text{ cal (pH 7.0)}$$

but when equivalent CO_2 is produced in a biological system (through a multistep reaction),

$$C_6H_{12}O_6 + 6O_2 + 38ADP + 38H_3PO_4 \rightarrow 6CO_2 + 38ATP + 44H_2O$$
$$-382,000 \text{ cal (pH 7.0)}$$

A noteworthy departure is the conservation of $-304,000$ cal/mol of glucose and gain of 38 moles of ATP, energy-rich (bond) compounds, i.e., 800 cal/mol of ATP. It also means 50,666 cal of energy are wasted if, on average, 1 mole of carbon dioxide produced chemically is wasted in the form of heat, an inferior quality of energy.

A simple calculation will reveal that each nutrient has some specified energy or calorific values. This can be compared to the different energy

TABLE 1.2 Comparison of Some Common Fuels

Fuel	Kcal	Thermal efficiency (%)	Effective heat (kcal)
Gobar gas, m^3	4713	60	2828
Kerosene, L	9122	50	4561
Firewood, kg	4708	17	800
Dry cow dung, kg	2092	11	230
Charcoal, kg	6930	28	1940
Soft coke, kg	6292	28	1762
LPG (butane), kg	10,882	60	6529
Coal gas, m^3	4004	60	2402
Electricity, kWh (hot plate)	860	70	602

SOURCE: Permission from KVTC, Mumbai.

values of different fuels, i.e., coal, kerosene, firewood, and so forth (see Table 1.2). Taking glucose as a model carbohydrate,

$$C_6H_{12}O_6 + 6O_2 \rightarrow 6CO_2 + 6H_2O - 686,000 \text{ cal}$$
$$(\text{molecular weight, MW} = 180 \text{ g}).$$

$$\frac{686,000 \text{ cal}}{180 \text{ g}} = 3800 \text{ cal/g}$$

and taking palmitic acid as model fatty acid,

$$C_{16}H_{32}O_2 + 23O_2 \rightarrow 16CO_2 + 16H_2O - 2,338,000 \text{ cal (MW} = 256 \text{ g)}$$

$$\frac{2,338,000 \text{ cal}}{256 \text{ g}} = 9133 \text{ cal/g}$$

Similarly, in amino acids, peptides show roughly the same value as that of carbohydrates. In biological systems (measurement through metabolic cage), it has been found that the biological energy values are slightly higher than those shown theoretically. This is more so by "specific dynamic action." When mixed foods particularly protein are taken, the total calorific value is enhanced. The exact reasons are not yet clear. Let us concentrate on a few examples in the following:

In ethanol fermentation (pH 7.0),

$$C_6H_{12}O_6 \rightarrow 2[C_2H_5OH + CO_2] - 56,000 \text{ cal}$$

In lactic fermentation,

$$C_6H_{12}O_6 \rightarrow 2[CH_3CHOHCOOH] - 47,000 \text{ cal}$$

But in lactic fermentation from polysaccharide,

$$(\text{Glucosyl})_n \rightarrow 2[\text{CH}_3\text{CHOHCOOH}] + (\text{Glucosyl})_{n-1} - 52{,}000 \text{ cal}$$

$$\text{CH}_3\text{CHOHCOOH} + 3\text{O}_2 \rightarrow 3\text{CO}_2 + 3\text{H}_2\text{O} - 319{,}500 \text{ cal}$$

If glucose is the starting point (as is the case of ethanol fermentation), then 2 moles of ATP are invested and finally 2×2 moles of ATP are regenerated and the net gain of ATP remains 2 (see Fig. 1.1). But if glycogen is the starting point, then only 1 mole is invested in the formation of fructose 1,6-diphosphate.

Hence, net gain in ATP is $4 - 1 = 3$. Twice a mole of reduced Co I is produced by the conversion of 3 phosphoglyceraldehyde to 1,3 diphosphoglycerate.

$$\text{ATP} + \text{H}_2\text{O} \rightarrow \text{ADP} + \text{H}_3\text{PO}_4 - 8000 \text{ cal}$$

But ΔF of formation of ATP $= +12{,}000$ cal.

The energy conservation or efficiency factor can be calculated in two different ways:

1. How much potential energy-rich chemical compounds are now gained?
   a. Ethanol fermentation: $-16{,}000/-56{,}000$, about 29%
   b. Lactic fermentation: $-24,000/-52{,}000$, about 46%

2. How much energy of reaction has been utilized as heat of formation of the energy-rich compounds?
   a. Ethanol fermentation: $24{,}000/-56{,}000$, about 43%
   b. Lactic fermentation: $36{,}000/-52{,}000$, about 69%

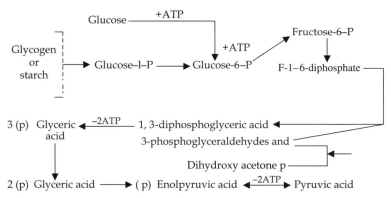

Figure 1.1 Anaerobic part of biological oxidation.

The percentage efficiency figures raise doubt about the interpretation. Such efficiency is never achieved by a man-made machine but biological systems can. If we accept the lower figures with a margin, we are conserving no less than 25% of our expenditure in the form of provident fund energy, even under sudden stress, i.e., anaerobic conditions.

Let us look at the situation when a reduced coenzyme is regenerated or oxidized (brief and simplified):

$$NADH\ (H^+) \xrightarrow{-ATP} FAD \xrightarrow{-ATP} Cytochrome \xrightarrow{-ATP} Cytochrome \xrightarrow{+\frac{1}{2}O_2} H_2O$$

Stoichiometrically,

$$CoIH(H^+) + \frac{1}{2} O_2 + 3ADP + 3H_2PO_4 \rightarrow CoI^+ + 3ATP + 4H_2O$$

Similarly in the oxidative part, through the tricarboxylic acid cycle, the major aspects may be represented as in Fig. 1.2.

From alpha ketogluterate to succinate, 1 mole of energy-rich phosphate in the form of guanosine triphosphate (GTP) is gained. Succinate to fumarate mediated by FAD coenzymes generates two equivalents of ATP. In the rest of the events, 4 sets of reduced Co I, when regenerated, give rise to $4 \times 3 = 12$ equivalents of ATP. In the entire sequence of events, from pyruvate plus oxaloacetate into citrate/isocitrate and finally back to oxaloacetate, a total of 15 equivalents of energy-rich phosphate bonds (ATP) are gained.

In combining the anaerobic part, 2 additional moles of reduced Co I will be reoxidized and 6 ATP equivalents will be regenerated. Starting from glucose-6-P all the way to CO_2 and H_2O, we see that $2 + 6 + (2 \times 15) = 38$ equivalents of ATP are gained. The balance of the equation has been

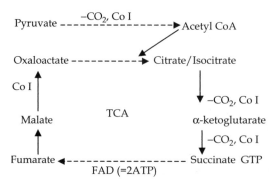

Figure 1.2 Tricarboxylic acid cycle (oxidative pathway).

cited earlier. An oxidative pathway is considered to be more effective from a biochemical energetic viewpoint.

One anabolic example of photosynthesis is briefly discussed. Theoretically, reversal of this known reaction should fit well for photosynthesis:

$$C_6H_{12}O_6 + 6O_2 \rightarrow 6(CO_2 + H_2O) - 686,000 \text{ cal}$$

But in fact, we find a slightly different figure. The entire reaction may be symbolically represented as

$$2H_2O + 2NADP^+ \xrightarrow[\text{Chloroplast}]{h\upsilon} 2NADPH \ (H^+) + O_2$$

$$3CO_2 + 9ATP + 5H_2O \longrightarrow \text{Triosephosphate} + 9ADP$$
$$+ 6NADPH \ (H^+) \qquad\qquad\qquad + 8H_3PO_4 + 6NADP^+$$

But the actual stoichiometric presentation shows

$$n(CO_2 + H_2O) \rightarrow (CH_2O)_n + nO_2 + n(113,000 \text{ cal})$$

almost 22,000 cal higher than expected; fortunately, however, the endergonic reaction derives its energy from light energy. These figures are justified because the part of the reaction occurring in the absence of light needs a large excess of energy-rich compounds (ATP). The deficiency of ATP is, however, taken care of by two linked reactions:

Cyclic photophosphorylation:

$$nADP + nH_3PO_4 \xrightarrow{h\upsilon} nATP + nH_2O$$

Noncyclic photophosphorylation:

$$4Fe_{ox} + 2ADP + 2H_3PO_4 + 4H_2O \xrightarrow{h\upsilon} 4Fe_{red} + 2ATP + O_2 + 2H_2O + 4H^+$$
$$\text{or } 2Co \ II_{red} + 2ATP + O_2$$
$$+ 2H_2O + 2H^+$$

The deficiency of 1 mole of ATP per mole of CO_2 fixed is provided by cyclic photophosphorylation. The other anabolic process is the nitrogen fixation, which is also highly energy consuming.

The heat of formation of NH_3 by a chemical pathway can only be determined indirectly. By the Haber process, high pressure and temperature is needed and the yield remains very low. So the input in energy in the technological process remains in large excess than the theoretical heat of formation of NH_3.

Nitrogen fixation can take place in nature in two major ways. Molecular nitrogen is converted to oxides of nitrogen in the atmosphere by electrical discharge and gets into soil by rainwater in the form of nitrites and nitrates. These are reduced to ammonia by the biological nitrogen fixation of symbiotic organisms or by blue-green algae.

In *Escherichia coli* and *Bacillus subtilis*, NO_3^- is reduced to NH_3

$$[NO^{-3} \rightarrow NO_2^{-1} \rightarrow N_2O_2^{-2} \rightarrow NH_2OH \rightarrow NH_3]$$

and an oxidation reduction potential of 0.96 V (pH 7.0) is utilized by these systems to convert other materials to a more oxidized state.

$$NH_3 + \frac{3}{2}O_2 \rightarrow NO_2^- + H_2O + H^+ - 36{,}500 \text{ cal}$$

$$NO_2^- + \frac{1}{2}O_2 \rightarrow NO_3^- - 17{,}500 \text{ cal}$$

$$N \equiv N \xrightarrow{2e^-} HN = NH \xrightarrow{2e^-} H_2NNH_2 \xrightarrow{2e^-} 2NH_3$$

Via Mo-protein complex

Hydrogen is made available from reduced coenzymes, and the energy is available from ATP produced by the oxidation of general metabolites.

In some systems, H_2 becomes the by-product, and this could be an ideal fuel or it can be used in a suitable chemical cell for the production of energy.

1.6 Chemical Cell

Two different metals in contact with a polar or ionic fluid generate the flow of electrons. When touched simultaneously by two different metallic rods, muscles contract, a pioneering observation that gave birth to the study of galvanic, voltaic, and Daniel cells.

The potential generated depends on the energy of sublimation, the ionization potential, the electronic work function, and the energy of solvation of ions. The nature of the solvent influences the last factor. The electronic work function also includes several other conditions of ionic activity. As a result, a potential difference will arise out of a simple concentration gradient, provided that anionic and cationic stoichiometry is maintained. A review of the existing knowledge is worthwhile here.

If two small baths, each having either Zn or Cu metal and corresponding dilute solutions of Zn^{2+} and Cu^{2+} salts, are in electrical continuity—say through a capillary of a U tube or a Pt wire—then current will flow in the two metals when connected outside, with Cu behaving as a cathode and Zn as an anode (see Fig. 1.3). The setup can also be designed by separating the two systems by a semipermeable membrane.

A similar experience is the cylindrical design of the commonly available dry cells, where a graphite rod at the center serves as a reference

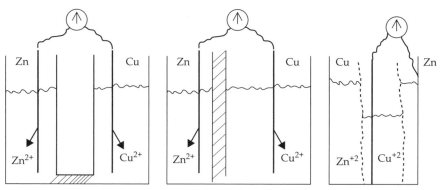

Figure 1.3 Chemical cell.

cathode surrounded by a paste of chemicals, usually NH_4Cl, totally housed in a small cylindrical cup of metallic Zn as an anode.

In each case, Zn gets oxidized and changes to Zn^{2+}, and Cu^{2+} is reduced and is deposited as Cu; in the graphite (carbon) electrode, the chemical change is not noticeable. (Theoretically, CH_4 should be formed, but slow escape of NH_3 takes place.)

The field of electrochemistry has progressed considerably. Standard electrode potentials and electrochemical charts with a fair degree of accuracy and reliability are available. Taking Pt (inert) electrodes, hydrogen gas at 1-atm pressure, immersed in a solution of hydrogen ion of unit activity is usually a reference or standard hydrogen electrode (usually referred as zero or standard scale). If an element goes into a solution, producing cation ($Zn \rightarrow Zn^{2+} = +0.761$ V), the half cell will give an oxidation potential with a sign opposite to the potential when the cation of the same species is deposited as the element, giving rise to a reduction ($Zn^{2+} \rightarrow Zn = -0.761$ V); the numerical values are expected to remain in the same order.

One may observe, on the other hand, that alkali metals have a tendency to become hydrated oxides in water, so they exhibit a tendency to offer oxidation potential with a + sign. When the element approaches nobility, then converts to the halogen ($2X^- \rightarrow X_2 + 2e^-$), the situation is reversed. A representative partial list of the standard electrode potentials is reproduced (see Table 1.3). So one may expect that in a chemical cell with $Zn/ZnCl_2$-$CuCl_2/Cu$, the EMF will be $+0.761 -(-0.340) = 1.101$ V.

If the electrode pair is made of the same material in a system, and the concentration difference of electrolyte is maintained between the two electrodes, a standard potential difference is expected, at the rate of 0.054 V per each tenfold rise in ionic concentration (referred to as concentration cells).

TABLE 1.3 Standard Electrode Potentials at 25°C

Electrode	Potential, V	Electrode	Potential, V
$Li \rightarrow Li^+$	+3.024	$Pt, \frac{1}{2}Br_2 \rightarrow Br^-$	1.087
$Na \rightarrow Na^+$	+2.715		
$Mg \rightarrow Mg^{2+}$	+2.34		
$AL \rightarrow AL^{3+}$	+1.67	$Pt, \frac{1}{2}Br_2(l) \rightarrow Br^-$	−1.065
$Zn \rightarrow Zn^{2+}$	+0.761		
$Fe \rightarrow Fe^{2+}$	+0.441	$Pt, \frac{1}{2}I_2(s) \rightarrow I^-$	−0.535
$Sn \rightarrow Sn^{2+}$	+0.140		
$Pb \rightarrow Pb^{2+}$	+0.126		
$Cu \rightarrow Cu^{2+}$	−0.340	$Pt, \frac{1}{2}O_2 \rightarrow OH^-$	−0.401
$Ag \rightarrow Ag^+$	−0.799		
$Hg \rightarrow Hg^{2+}$	−0.799	$Pt, Fe^{2+} \rightarrow Fe^{3+}$	−0.771
$Au \rightarrow Ag^{3+}$	−1.300	$Pt, Pb^{2+} \rightarrow Pb^{4+}$	−1.75
		$Pt, Sn^{2+} \rightarrow Sn^{4+}$	−1.75
$Pt, \frac{1}{2}H_2 \rightarrow H^{\pm}$	±0.00	$Pt, Cu^+ \rightarrow Cu^{2+}$	0.16
$Pt, \frac{1}{2}Cl_2 \rightarrow Cl^-$	−1.358	$Pt, \frac{1}{2}Hg^{2+} \rightarrow Hg^{2+}$	0.91

If Zn is used as a common electrode, or better inert-metal electrodes are used (e.g., Pt) and immersed into NH_4Cl or HCl solutions, say 0.1 and 1.0 N, a potential difference of 0.054 V will be experienced. The effect of temperature and other factors which affect ionic activity will definitely alter the values of EMF. The strength of the current will depend, expectedly, on the total surface area or participation of the total number of ions and their charge-carrying capacities.

Electrochemical behavior of certain elements, e.g., carbon and silicon, must be determined indirectly. Only graphite exhibits direct application in a chemical cell, but other forms of carbon or silicon do not play any significant role at this state of knowledge (see Fig. 1.4).

1.7 Models of Bioenergy Cells

One attractive suggestion is based on harvesting the potential produced in different steps of metabolism in living systems [2]. Basic principles remain the same in all such models. One of them is to tap the oxidative phosphorylation path, and the other one is to use the photosynthetic mechanism. There are a few more novel systems suggested by other schools: (a) calcium pumps in biological systems by Ernesto Carafoli of Swiss Federal Institute of Technology, Zurich, (b) constructing cells from bacteriorhodopsin of the purple membranes of certain bacteria by Lester Packer of the University of California at Berkeley, United States, and

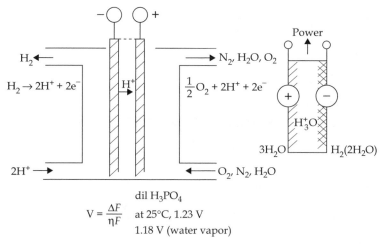

Figure 1.4 Gaseous battery (hydrogen fuel cell).

(c) isolated energy-rich compounds, i.e., iron-sulfur proteins, ATP, and so forth, suggested by many other authors [3].

1.7.1 Oxidative phosphorylation path

In the electron transfer chain, the conversion takes place at lower potentials, i.e., NAD/NAD^+ to $NADH/NADH^+$ between ± 0.6 V, favorable at a pH higher than 7.0. But the process develops other energy-rich compounds, and thus, very little free energy in the form of heat is directly available.

1.7.2 Photosynthetic path

Cytosolic to mitochondrial compartments, the interconversions of pyruvate to aspartate and to glutamate; malate to α-ketoglutarate; the energy produced is utilized to synthesize higher carbon compounds, ultimately to glucose or even polysaccharide and polynucleotide (genetic material) (see Fig. 1.5). Artificial culture of thylakoid or chloroplast, (only remains a possibility for academic purposes at present); cannot be commercially achieved as yet.

The most important achievement is the photolysis of water (see Fig. 1.6), i.e., production of proton to hydrogen, reduction of carbon dioxide, reduction of nitrogenous material, and increase in nitrogenous and carbonaceous biomass. Attempts have been made to utilize the energy-trapping process of the photosynthetic pigments of the plastoquinones at two stages: (1) Pigment II utilizes 680–700 nm, converts water to a more

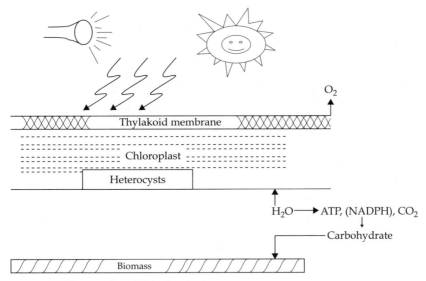

Figure 1.5 Electron flow in biophotolysis.

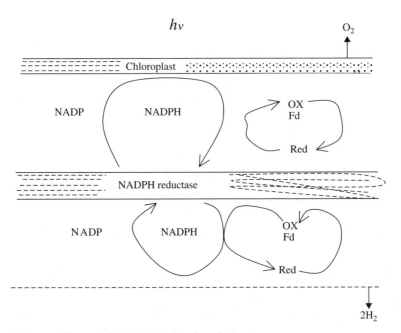

Figure 1.6 Separated photolytic chamber design.

energetic intermediate, and undergoes a change of $+0.8$ to -1.1 V. (2) Pigment I utilizes 700–730 nm, undergoing $+0.5$ to almost -1.4 V, production of hydrogen, oxidation of coenzyme, making electrons available.

Models can be created where direct tapping from the thylakoid membrane may be made possible. Electrochemical cells have been designed where living thylakoids are used and exposed to sunlight from which, through proper instrumentation, the energy can be tapped.

1.8   A Living Cell Is an Ideal Cell

Quite a few prototype experiments have been done, and a large number of postulations are yet to be worked out, based on the potential difference maintained within and outside the living cell. Two well-known phenomena are the membrane potential and the injury potential.

If the inside and outside walls of a cell membrane are brought to electrical continuity, current will flow. Usually the inside is anodic, mainly due to the dominating fixed charges on the membrane protein. When injury is caused, the excess mobile cations from the outer surface infiltrate the inner layer and a local flow of current takes place. A healthy (uninjured) cell maintains an intact membrane, spends some metabolic energy to pump in nutrients and K^+, and retains them within the cell against a concentration gradient. Likewise, some of the metabolic products, including Na^+ are pumped out (exceptions, namely, Halobacterium— are few).

Most of these functions are chemically mediated (by ATPase, ATP – Mg^{2+}, etc.) and amount to mechanical work. Maintenance of the potential difference on the membrane inside out is an indirect electrical manifestation of the chemical activity. The membrane components, particularly protein, uphold its configuration with desired functional groups projected within. Retention of selective ions with the cell, in addition to offering electrical neutrality, offers colloid osmotic steady state (through Donnan equilibrium).

Another interesting phenomenon associated with chemical activity of cells is the pH specificity of specialized cells. Normally, the mammalian body fluid behaves as an alkaline buffer, pH 7.4, with only about 0.1 M, contributed by metal ions, but has high osmolarity due to colloid osmotic components. In spite of the pH 7.4 of the circulating fluid, the stomach, part of the kidney, and the respiratory system maintain distinct acid pH. This mechanism of upholding higher H^+ concentration is by metabolic expenses. In plants, the tissue fluid is usually acidic, say pH 6.5, and certain specialized tissues, namely fruits, exhibit strong acidity. In very rare cases (marine flora), plant tissue fluids show alkaline pH.

These examples are sufficient to indicate that if gastric mucosa is connected to the intravenous system, a potential difference or an EMF will be experienced. Likewise, if the root tissue and the fruit of a tree are short-circuited, current (however feeble) will be experienced. This information is not worth much at this present state of the art because the magnitude of instrumentation will appear prohibitive. But in space research, there was no alternative left but to develop solar cells, and silicon cells have found their place despite their cost. Because roughly 4 kcal of energy is available per gram of coal or hydrocarbon, this technique is of limited value at present. However, with enhanced improvement, the renewable resources of flora and fauna may be sources of direct energy when we run out of oil and coal and will also appear inexpensive under those circumstances.

1.9   Plant Cells Are Unique

Whether they are green algae (chlorella) or the higher plants, autotrophs in general are gifted in nature to fix carbon dioxide and produce biomass. In ecologic terms, these are producers. The dominant autotroph is phototrophic. Photosynthesis has two distinct aspects: the light dependent step, where photolysis of water takes place:

$$NADP^+ + H_2O \xrightarrow{\text{680 nm (52 kcal)}} H^+ + NADPH + \frac{1}{2}O_2$$

$$ADP + P_i \xrightarrow{\text{8 kcal}} ATP$$

During this reaction, oxygen is set free, Co II is reduced and phosphorylation of ATP takes place. The photoenergy is chemically utilized twofold.

In the next step, through a very complex enzymic sequence, CO_2 is incorporated into the existing metabolite pool and higher carbohydrates are biosynthesized. This step of the reaction finds variation in different species; carbohydrates, proteins, and lipids are biosynthesized. Then, the first part of the reaction makes the autotrophs unique. Light falling on chloroplasts develops an electrical field across the membrane.

In the presence of the pigments in chloroplasts, the light energy is trapped and activates water and lyses it. Ideally, water, if converted into its elemental components, requires (at 25°C) 68.3 kcal/mol (from liquid) or 57.8 kcal/mol (from vapor). Thermal energy is not sufficient to bring this change. During photolysis, the plant pigment augments electron flow, and the electron flow system culminates in the two energy-rich chemical products (reduced Co II, ATP, and O_2), as already mentioned (see Fig. 1.7).

Lester Packer's group at the University of California at Berkeley has shown the steps of the pathway with chloroplasts from spinach leaves,

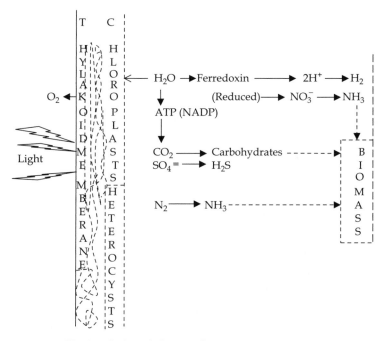

Figure 1.7 Biophotolysis and electron flow system.

ferredoxin from *Spirulina,* and hydrogenase of *Clostridium pasteuri-anum* [3].

$$2H_2O \rightarrow 4H^+ + 4e^- + O_2$$

$$4 \text{ Ferredoxin}^+ + 4e^- \rightarrow 4 \text{ Ferredoxin}$$

$$4H^+ + 4 \text{ Ferredoxin} \rightarrow 2H_2 + 4 \text{ Ferredoxin}^+$$

$$O_2 + \text{Glucose} \rightarrow \text{Gluconate} + H_2O_2$$

$$H_2O_2 + \text{Ethanol} \rightarrow 2H_2O + \text{Acetaldehyde}$$

The overall reaction is

$$\text{Glucose} + \text{Ethanol} \rightarrow \text{Gluconate} + \text{Acetaldehyde} + H_2$$

Two H_2 are produced for each O_2 produced (if not consumed by an oxidase-type reaction as shown previously). Dr John Benemann of the same university has also suggested that hydrogen and methane production is possible by designing a two-stage system separated from each other (see Fig. 1.8).

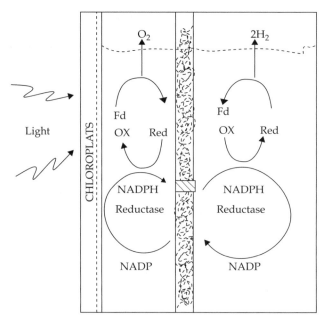

Figure 1.8 Concept of a two-stage separated system for photolytic chamber.

Dibromothymoquinone blocks the natural electron flow system at plastocyanin level (see Fig. 1.9). Thus, in the presence of an artificial donor or acceptor, the photo systems I and II can be separated at pre- and post-blocking points.

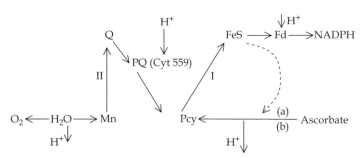

(a) Phenazin methesulphate

(b) Diaminodurol; I and II photosynthesis systems I and II

Figure 1.9 Electron flow system at the plastocyanin level.

1.9.1 Photosynthetic bacteria

Small vesicles, called chromatophores, can be isolated from the membranes of photosynthetic bacteria, which exhibit two types of electron transfer chains resembling mitochondria and chloroplasts. Chroma-tophores supported on artificial membranes permit the generation of 200 mV on illumination. The salt-bacteria (Halobacterium halobium) contain a simple protein–vitamin A aldehyde, known as bacteriorhodopsin, when supported on artificial membranes that generate 250 mV on illumination. This system is simpler than its counterpart. There is a probability that the entire system may be successfully synthesized or assembled. Solar photocells made of bacteriorhodopsin show great promise.

1.10 Biofuels

Prospects of ethanol and biodiesel as substitutes for conventional fuels will not be discussed here; these two aspects are presented in sufficient detail in Chaps. 3, 4, 5, and 6. One of the promising approaches for future fuel is, perhaps, hydrogen and methane, both of which could be obtained from living, particularly microbial resources.

Photosynthesis is the main route through which oxidized carbon is reduced and again oxidized back to carbon dioxide for the generation of energy. Based on this principle, we can utilize a few steps from this life chain. This topic could be called biophotolysis—alternatively, photobiolysis.

In the system, direct electron transport from water to hydrogen has not been demonstrated as a technically feasible reaction. For this, continued research is required to elucidate the basic nature of FeS (PEA, ferredoxin, and hydrogenase). This may lead ultimately to the practical feasibility of production of hydrogen (ideally 20 μL/h). Section 1.16 discusses hydrogen in detail. One inherent problem is the stability of the hydrogenase system because of its sensitivity to molecular oxygen produced during photosynthesis.

However, one may design a two-step or two-compartment system. Reduced Co II is the oxygen-stable electron carrier between photosynthesis and hydrogenase. A higher ratio of reduced Co II or Co II helps the evolution of hydrogen, in spite of the unfavorable redox potential of the coenzyme. Only Co II (reduced) can be pumped or transported from one stage (compartment) to the other. Photosynthesis and hydrogenase systems have to be encapsulated or immobilized separately in order to retain their respective activity; the two stages or compartments may be connected through fiber filters. An example could be to use appropriate algae to produce reduced organic compounds which can be pumped into bath of photosynthetic bacteria of hydrogen fermentation.

One partial modification will be to collect oxygen during the day and hydrogen at night, at the expense of accumulated reduced coenzymes, made operative by anaerobically adapted microalgae or nonheterocystous nitrogen- fixing blue-green algae. For product separation, the enzyme technology or immobilization is inapplicable for biophotolysis. However, there are potential practical applications of immobilized hydrogenase in biochemical hydrogen–oxygen fuel cells. If such enzymes can be immobilized on an electrode surface, an inexpensive fuel cell might be developed, which would increase the energy recoverable for hydrogen to save fuels.

Awareness of the limitations due to efficiency, engineering, and the economy of these principles will save disappointment and encourage continued research. Geographical location and frequency of weather change limits the insolation. The best photosynthetic efficiency is only 6% of the total incident solar radiation, i.e., 5 kg/(m$^2 \cdot$ yr) of H$_2$ by biophotolysis. Half of this could be a very satisfactory achievement.

1.10.1 Heterocystous blue-green algae (example, *Anabaena cylindrica*)

The heterocyst, regularly spread among more numerous vegetative cells (ratio 1:15), receives carbon compounds fixed by the neighboring vegetative cells in exchange of the nitrogenous compounds fixed by them. Nitrogenase, like hydrogenase, needs an anaerobic environment to function and can produce hydrogen only under certain conditions (absence of molecular nitrogen). The ratio of evolution of hydrogen and oxygen roughly corresponds to the ratio of the heterocysts and vegetative cells and also with the ratio of nitrogen and carbon for nutritive requirements.

If the algal culture is exposed to argon for about 24 hours, due to nitrogen starvation, differentiation of the heterocysts increases from 6% up to 20%. In addition, a yellowish color appears due to the loss of the light-trapping pigment phytocyanin, resulting in less carbon dioxide fixation, i.e., oxygen evolution and an increase in light conversion efficiency by almost 0.5%. Induction of reversible hydrogenase in the heterocysts, as its theoretically higher turnover principle, is less affected by N$_2$ and O$_2$, and independent of ATP, it becomes more desirable and needs heterocysts to be genetically improved.

1.10.2 Photofermentation by photosynthetic bacteria (example, *Rhodospirillium rubrum*)

Hydrogen production by photoheterotrophic bacteria is principally similar to that of blue-green algae, capable of fixing nitrogen and producing hydrogen. The microbes are capable of converting large varieties of organic compounds to carbon dioxide and hydrogen up to 50 kg/(m$^2 \cdot$ yr). Practical applications of these bacteria are more of an engineering problem than one of scientific "know-how." The scope of newer research exists on the noncyclic hydrogen production by these microbes unin-hibited by nitrogen. Dilute wastes can be utilized by the photosynthetic bacteria, which is an added advantage over those of the methane fer-mentors. The conventional fermentation of organic substrates to methane or hydrogen is theoretically limited to 80% and 20%, and prac-tically to 65% and 15%. The difference is accounted for by the synthesis of ATP and cell biomass. ATP is produced in presence of light and reac-tions are driven at its expense, if hydrogen is produced by nitrogenase.

1.10.3   Methane production

The biology of an "oxidation pond" is not well understood. The algae-versus-bacterial growth needs to be controlled, and anaerobicity and temperature need to be maintained properly. The carbonaceous matter tends to ferment, and methane is produced instead of carbon dioxide. The end product, methane, can be used either as a direct fuel or through a suitably designed fuel cell. Microbial methane and hydrogen production are discussed later.

1.11   Plant Hydrocarbons

While a significant number of scientists are assessing the future of renewable and nonrenewable sources of energy, and their potential use-fulness and costs, a few of them are busy exploring existing storehouses of nature and modifying the renewable resources into direct conventional fuels. Prof. Melvin Calvin and his group at the University of California at Berkeley emphasize the importance of a group of plants which, in addi-tion to producing polysaccharide, also produce polyisoprenes (rubber) and similar associated products [4]. While the *Hevea* produces rubber, different euphorbiacea produce polyhydrocarbons that have molecular weights lower than 10% of that of average natural rubber. It is likely that chemical manipulation may yield liquid fuels similar to that of conventional gasoline or diesel out of these products.

The interesting aspect of these plants is that rubber plants demand good insolation and high moisture content in soil as well as in the atmos-phere. But many subspecies of *Euphorbia* can grow comfortably in sunny semiarid lands, where standard cultivations are not economically viable [5]. This leads us to two major considerations: (1) soil conservation, eco-logic improvement, and increase in P/R (productivity/respiratory) ratio; (2) production of hydrocarbon and biomass, both of which have energy value.

Avalois is the North Brazil variety, and *Euphorbia tirucalli* is the Southern Californian equivalent of the plant. Both of them usually con-tain 30% hydrocarbon in their latex. Similar or parallel plants in the Indian Subcontinent are not yet well known. But like rubber plantation, which successfully migrated from Brazil to Malaysia, one may try a few

TABLE 1.4   Yield of Some Important Crops and Their Biomass Utilization

	Approximate composition (%)			
Latex	Moisture	Water sol.	Organic matter high MW	Organic matter Low MW
Heve	65	3	31	1
Euphorbia	63	9	27	1

Ideal yield of some crops	
Crop	MT/ Ha/ Yr
Sugarcane	30
Sugar beet	33
Algae	87
Sorghum	36
Corn	13
Eucalyptus	54
Rubber (Malaysia)	2

Example of chemical diversification of biomass

Sugar cane	⟶	Cane Juice	⟶	Ethanol
↓		↓		↓
Bagasse		Butanol		Ethylene
(Cellulose Lignin)		Glycerol		Chemicals
Fodder		Citric Acid		Ethyl chloride, etc.
Energy		Aconitic Acid		

species of *Euphorbia*—particularly on the rocky, arid, or laterite belts, which are rather unproductive for forestry or cultivation. It is worthwhile to take a glance at some information already available on these products [6].

1.12   Biogas

Age-old phenomena of spontaneous combustion of natural gas, continuously or intermittently, were called "will-o-wisp" or "fool's fire." Later, these phenomena were assigned to "marsh gas" and mainly methane by H. Tappeiner (1882) [7]. Almost a century passed, through which different postulates had to be verified in order to unveil the mechanism behind this natural methanogenesis or biogas formation. First, one-step microbial degradation of cellulose to methane was proposed. This was replaced by a two-step concept, where lower-molecular-weight organic acids are produced as intermediates, which further undergo conversion to methane. Finally, the three-step concept has been prevailing (the entire process is anoxic):

$$\begin{array}{ccccc} \text{Hydrolytic} & & \text{Acetogenic stage} & & \text{Methane, organic} \\ \text{fermentive stage} & \rightarrow & \text{(Mesophilic)} & \rightarrow & \text{(Thermophilic)} \end{array}$$

$$\text{Organic matter} \quad \rightarrow \quad \begin{array}{c} \text{Organic matter} \\ \text{Alcohols, } H_2,\ CO_2 \end{array} \xrightarrow{\text{(35°C, pH 5–6)}} \begin{array}{c} \text{Acetic acid} \\ H_2, CO_2 \end{array} \Big\downarrow$$

$$\text{Methane} \xleftarrow{\text{(45°C, pH 4–6)}}$$

$$CO_2$$

An oversimplified mass balance may be written as

$$C_6H_{12}O_6 \rightarrow 3CH_4 + 3CO_2$$

The technical values of yield coefficient, biological efficiency, chemical/biological oxygen demand (COD/BOD), biological efficiency in productivity/ecologic efficiency rate (BEP/EER) ratios, and so forth are yet to be established for each setup or system. Mostly obligate anaerobes and a few facultative microbes contributing to these conversions belong to different genera. A few may be mentioned: *Actinomyces, Aerobacter, Aeromonas, Arthrobacter, Bacillus, Bacteroides, Cellulomonas, Citrobacter, Clostridium, Corynebacterium, Enterobacter, Escherichia, Klebsiella, Lactobacillus, Laptospira, Micrococcus, Nocardia, Peptococeus, Proteus, Pseudomonas, Ruminococcus, Sarcina, Staphylococcus, Streptococcus, Streptomyces,* and many others. A few methanogenic species are also known: *Methanobacterium bryantii, Methanococcus vanniellii, Methano-genum aggregans, Methanomicro-bium mobile, Methanosarcina barkeri, Methanothrix concillii,* usually eukaryotic organisms, and blue-green algae are incapable of performing such bioconversions [8].

Morphologically, the organisms belong to wide groups: coccus, sarcina (flower-like), rod, filamentous, and other shapes. G + C (guanine-cytosine) values of DNA of these organisms also suggest that they all have varied origin and hence are likely to have different metabolic patterns. Khan (1980) found that *Acetivibrio cellulolyticus* producing acetic acid and hydrogen from cellulose are readily utilized by *M. Barkeri* to produce methane and carbon dioxide. It has been established beyond doubt that the process is chemolithotrophic metabolism, favored by strict anaerobic condition, and facilitated by the absence of sulfates, abundance of moisture, approximate temperature range of 25–40°C (37°C), and pH 6.2–8.0 (pH 6.8). The organic materials on which these organisms survive and grow are usually cellulose in nature. Crop residues, agricultural residues, animal excreta, municipal sewage, and other organic materials derived from terrestrial and aquatic origin are also considered as good substrates. Plant materials with high lignin content are an inferior type of feed for such reactions.

A pretreatment or partial putrefaction or degradation makes the process easy. In this respect, animal excreta appear to be a ready-made substrate. The art of producing gaseous fuel out of cattle excreta is well known in the Indian Subcontinent as the *gobargas plant,* and will be discussed subsequently.

Sargassum tenerrimum, an abundant variety of marine algae found on the Indian coast of the Arabian Sea, shows promising results in laboratory experiments by anaerobic digestion. A mixed culture of marine bacteria and methanogens happens to be a better choice. In a prototype experiment, the partially treated marine algal biomass mixed with cattle dung could be the initial feed for a digester. In a mixed culture, the entire process is a complex one. The organisms which are very efficient in cellulolytic activities degrade higher-carbohydrate materials into simpler products as lower organic acids, including CO_2 and less frequently H_2, along with other products, but very seldom show a significant amount of reduction reactions. In absence of methanogens, they usually produce H_2, CO_2 (even CO), formate, acetate, and less favorably other fatty acids and alcohols. It has been established that many methanogens utilize NH_4^+ as their nitrogen source, either H_2S or cysteine for their sulfur requirement, and other growth-stimulating amino acids, vitamins, and some trace minerals.

Uncommon in many other anaerobic organisms, methanogens have shown presence of a cofactor (coenzyme) named CoM, identified as $HSCH_2CH_2SO_3$ (2-mercapto-ethanesulfonic acid), and also another low-molecular-weight factor called F_{420}, as of yet unidentified. This F_{420} in an oxidized state fluoresces at 420 nm but loses all optical activity when reduced. This compound is neither a ferredoxin nor can it be substituted by ferredoxin. Another interesting part is its dependence on Co II (NADP) and it cannot be substituted by Co I (NAD system). Occurrence of oxidative or substrate-level phosphorylation in methanogens could not be established, and the presence of quinines or cytochrome b/c systems could not be observed. The involvement of methylcobalamin also could not be substantiated. So, a large part of the information is yet to be derived by the next-generation scientists. It will be useful to summarize some of the metabolic steps, so far understood (see Fig. 1.10).

The ecologic role of biogas is manifold. Chemical anoxic transformation reduces the BOD value of the organic residues, which in turn are enriched, proportionately in its C, N, P, and mineral ratios. In lignocellulosics, after the anoxic process, enrichment of lignin occurs and may lead to **peat formation**. This may be the origin of coal; natural gas and coal deposits are likely to be found within a reasonable stretch. This is a built-in machinery of nature for BOD and pollution control.

1.13 Gobargas

As already mentioned in the preceeding section of biogas, gobargas is an extended version of the biogas. Usually, when cattle excreta (gobar)

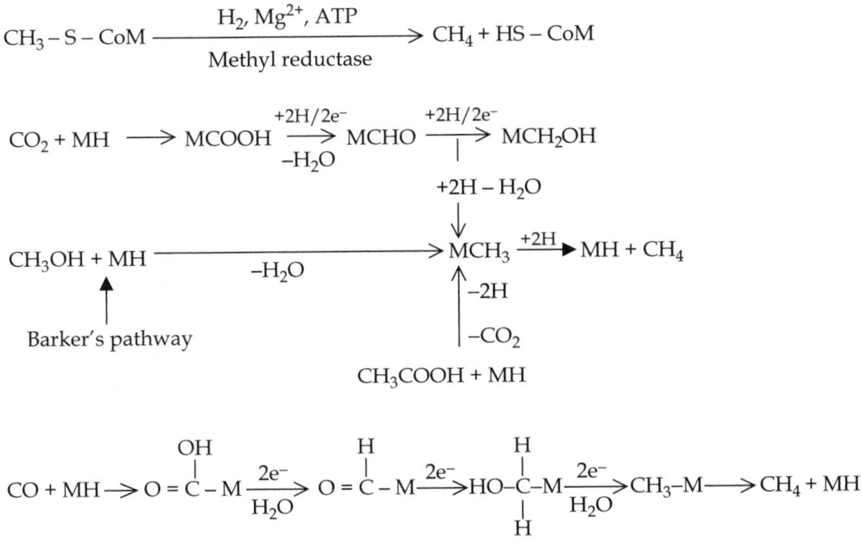

$$CH_3 - S - CoM \xrightarrow[\text{Methyl reductase}]{H_2,\ Mg^{2+},\ ATP} CH_4 + HS - CoM$$

$$CO_2 + MH \longrightarrow MCOOH \xrightarrow[-H_2O]{+2H/2e^-} MCHO \xrightarrow{+2H/2e^-} MCH_2OH$$

$$+2H - H_2O$$

$$CH_3OH + MH \xrightarrow[-H_2O]{} MCH_3 \xrightarrow{+2H} MH + CH_4$$

Barker's pathway

$$-2H$$

$$-CO_2$$

$$CH_3COOH + MH$$

$$CO + MH \rightarrow \underset{\substack{\parallel \\ O}}{O = C - M} \xrightarrow[H_2O]{2e^-} \underset{\substack{\mid \\ H}}{O = C - M} \xrightarrow{2e^-} \underset{\substack{\mid \\ H}}{HO - C - M} \xrightarrow[H_2O]{2e^-} CH_3 - M \longrightarrow CH_4 + MH$$

Gunsalus pathway

MH (reduced metabolite/reduced coenzyme/reduced enzyme complex)

Figure 1.10 Methanation.

is the starting material for anoxic fermentation to flammable gas, it is called gobargas. Before a scientific and technical approach was given to this promising field, the technique was developed in the southern part of India in a very crude way. Partly dehydrated animal excreta, when ignited, produces fumes and burn for a short duration with a partially sooty flame a little above the solid fuel. Slurried excreta, when stored in closed earthen vessels for a while, produced flammable gas. Based on these observations, villagers developed techniques of producing gas similar to illicit brewing.

Perhaps the greatest benefits of gobargas projects are secondary in nature. It takes out the pollution and ecologic problems and yields better biomass as compost and manure. The primary product, the biogas, has of course become very important in the present energy perspective. The fuel value of the gas, though not very high, is relatively safe and pollution free. Out of the many reports available so far, the positive and encouraging points leading to successful implementation of gobargas projects are very restricted. The negative points or factors which make the progress slow down are many, and a few are difficult to overcome. It may be useful to mention a few of them. These points are by no means insurmountable, but may help us to orient our future course of action, research, and development.

1. Dehydrated cow dung is a popular fuel and does not need special or expensive containers for keeping throughout the year.

2. Untended herds make the collection of dung laborious and cost intensive.

3. Installation of community biogas plants is not easy. Due to the frag-mentized small households, individual plants are also difficult to erect. Most families cannot provide the minimum 50-kg average dung input to the plant. About 50 L of water should also go with it. Fifty percent of the settlements are located in drought-prone areas. The remaining 50% face water shortage during the 5 months of dry season.

4. Temperature fluctuations throughout the year are significant and affect the rate of biogas production.

Disfunctioning and malfunctioning of some of the plants, due to the lack of proper maintenance and servicing, create poor examples to neigh-bors. This reduces the fresh installation potentialities and leads to an unwillingness to invest funds. The increasing cost of installation is another reason for the negative attitude.

The Chinese use mostly underground designs, and their outlays have been more successful because they have already undergone a genera-tion of restructured social order. As per Neelakantan's (1974–1975) report, the wet-dung yield of a cow is on an average 11.3 kg (3.6 to 18.6 kg) and of a buffalo is 11.6 kg (5.0 to 19.4 kg). The daily output of dung from an average of five cattle (a minimum of four) may suffice for a house-hold with a miniature gobargas plant. When underground ambient con-ditions (30°C), are favorable, at least 2.7 m^3 of gas (50 m^3/ton of wet dung) per day is expected out of the plant. This gas has a minimum of 9500 kcal (3500 kcal/m^3) of heat value (equivalent to 1.5 L of kerosene), which may serve the daily need of a five-member family. It is estimated that the average daily requirements of the gas per adult per day are 0.3 m^3 for cooking and 0.2 m^3 for lighting purposes.

Installation of a 3 m^3 digester (gobargas plant), partly embedded in the earth, or preferably constructed underground, as per improved versions of several designs, suffices for one standard household (see Fig. 1.11). At the present cost, it comes to about Rs. 10,000 (approximately US $200), depending on the remoteness of the house or the community. Attractive cost figures have been developed by competent engineers and social workers who have estimated an annual savings to the tune of Rs. 1000 (approximately US $20) per family, and the initial investment is likely to be paid off within 3 years. The estimated average lifetime of a gobargas

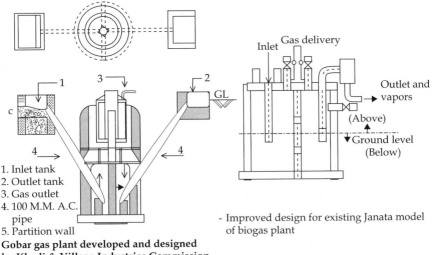

1. Inlet tank
2. Outlet tank
3. Gas outlet
4. 100 M.M. A.C.
   pipe
5. Partition wall

Gobar gas plant developed and designed
by Khadi & Village Industries Commission

- Improved design for existing Janata model
  of biogas plant

Figure 1.11 Gobargas plant.

plant is supposed to be 20 years. It is perhaps very important that a semi-skilled person or a trained "know-how" person tend to the plant.

Once installed, a 3 m^3 digester plant will require about 50–60 kg (4 buckets) of raw wet cattle dung and an equal amount of water. If the dung is slurried prior to feeding the digester plant, stirring may not be needed. Initially, a 15-day incubation is necessary and combustible gas starts coming out after about 3 weeks, when stabilized, and will continue to produce a gas mixture which is satisfactorily flammable. The average retention time of the materials in the digester is 3–7 weeks (average 5 weeks). The optimal temperature, of course, is 40°C (15–65°C) with a pH 6.8 (pH 6.5–7.5). In a small digester (family unit), control of temperature and pH remains out of bounds for ordinary villagers.

The omnipresent microbial flora in the ruminants will start the reaction, initially at a slow rate. No additional microbial culture is usually required. The gas is composed mainly of CO_2 and methane, and traces of other gases. Objectionable or harmful gases are very rare. Since a mixture of carbon dioxide is present, the gas is less flammable and hazardous than LPG, but needs sufficient precaution to be handled in the household. Most of the precautions to be observed in handling and using bottled gas will also apply in this case. The pipeline from the plant to the burner needs to be checked occasionally for leaks.

Human excreta and other animal excreta are equally useful for the same purpose. In fact, all such domestic excreta and pulped organic refuge may be mixed together to enrich the feed to the gobargas plant. Social practices and inhibitions prevent people from combining the feedstock materials. The common septic tank system can also be modified in design and be made to deliver biogas. The quantity of human excreta per family is relatively small, and hence, the gas evolved will hardly meet even the partial requirement of the family, if the biogas plant is fed exclusively with night soil.

The disappearing forests and forage have a cyclic relation in the ecosystem. Rising cost of animal feed of all kinds adds to the crisis. Keeping of cattle in small village households may not be an attractive proposal very soon. A major part of the animal dung is not collected by the owner of the cattle while animals graze. The space required to keep cattle and have a biogas plant will be considered a poor investment, due to soaring price of land, even in remote villages. Considering these and a few more unforeseen factors, better prospects of gobargas plants in a distant future may not be a correct speculation.

1.14   Biomass, Gasification, and Pyrolysis

1.14.1   Biomass

Imitating the coal-based process, biomass conversion has also been tried and looks promising. Main sources of biomass are agricultural, horticultural, and forest wastes. Municipal organic solid wastes (which are also plenty) are potential resources as well. Considering biomass as a renewable resource, the bioconversion may be pyrolytic, where biogas and bio-oil are the main products and yet the residue contains some calorie value which can be further utilized (as adsorbents, filter beds, chars, etc.). Supercritical conversion and superheated steam reformation of biomass are recent techniques. During 1990–1997, quite a few reports appeared in the literature showing success and promise of catalytic or uncatalytic reformation of biomass to hydrogen (almost to 18% v/v) without any char or residues.

Temperature ranges of 340–650°C, with pressures of 22–35 MPa, are cited with as low as 30-s residence time, through supercritical flow reactors. The raw materials are widely varying: water hyacinth, algae, bagasse, whole biomass, sewage sludge, sawdust, and other effluents rich in organic matters. In some efficient carbon bed–catalyzed reactors, other products (i.e., carbon mono- and dioxides and methane) were also detected.

1.14.2   Gasification and pyrolysis

Gasification, an exothermic reaction, yields mostly producer gas, a mixture of carbon monoxide, hydrogen, and methane at temperatures

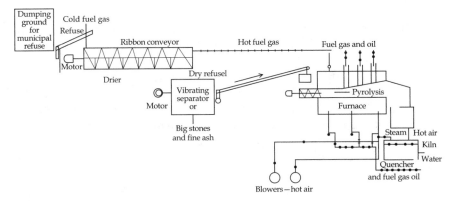

Figure 1.12 Flowchart of refuse processing plant.

above 1000°C, mostly in the absence of air. The starting materials may be any kind of organic matter, preferably waste materials like cotton and jute sticks, corn cobs, bagasse, and many other plant and vegetation products. In India, annually 16 million tons of rice husk, 160 million tons of paddy straw, 2 million tons of jute sticks, and 2.2 million tons of groundnut shells are available as agricultural by-products.

The gas can be directly used as fuel or used to drive irrigation pump sets. Several designs are available.

Pyrolysis, a thermochemical conversion, also performed in absence of air at a temperature of 500–600°C, yields gaseous components, hydrocarbons, carbon monoxide, hydrogen, methane, butane, some liquids, tars, and a little coke, all of which have very high energy content. Starting materials are similar to those mentioned under gasification. The vegetable matter in the municipal refuse (as much as 50%) is also good feed for pyrolysis. Very optimistic economic analysis for the pyrolytic process has been put forward by investigators, and a properly designed plant, say capable of handling 250 tons of organic refuse per day, will be fully paid off at the end of 5 years. There are 20 domestic or family-size models suggested by organizations. As per the available information, large-scale use of either gasifier or pyrolyser has not been noticed so far. But for the municipalities, the responsibility of quick disposal of the refuse and the environmental issues will prompt installation of such plants in the near future. One such flowchart of a model plant is given in Fig. 1.12.

1.15 Bioluminescence

A typical natural phenomenon, probably a unique mating signal by the "firefly," also exists in other living species, namely, bacteria, protozoa, fungi, and worms, in the forms that emit visible light. In most cases, the nature of the luminescent light varies in color and intensity; but chemical pathways are, to a great extent, common. The chemical products responsible for giving out different colors are different and are not yet fully known.

A heat-labile simple protein enzyme luciferase (MW 10^5) makes a complex (luciferyl adenylate E) with reduced luciferin, in the presence of ATP (Mg^{2+}), which subsequently breaks down into different products in the presence of molecular oxygen. This results in the excitation of luciferin to a high-energy state. On return of the same to the ground state, emission of visible light produces bioluminescence (see Fig. 1.13).

$$LH_2 + ATP\ (Mg^{2+}) + E \rightarrow LH_2 - AMP - E + PPi$$

$$LH_2 - AMP - E + O_2 \rightarrow Products + Light$$

The phenomenon appears to be insignificant but a substantial supply of luciferin, ATP (Mg^{2+}), and a little enzyme can deliver an appreciable luminescence of practical use. Whether luciferin, luciferase, and ATP may also be harvested from animal resources, or the chemical components may be synthesized economically and the enzyme can be procured from flies, remains a matter of investigation and development. Like bee-keeping, culture of "fireflies" is very likely to become a profitable art. The dream of producing high voltage by animal tissues, imitating the electric eel, may come true in the near future; the fundamentals are known, but economic viability is not assured, hence not discussed here.

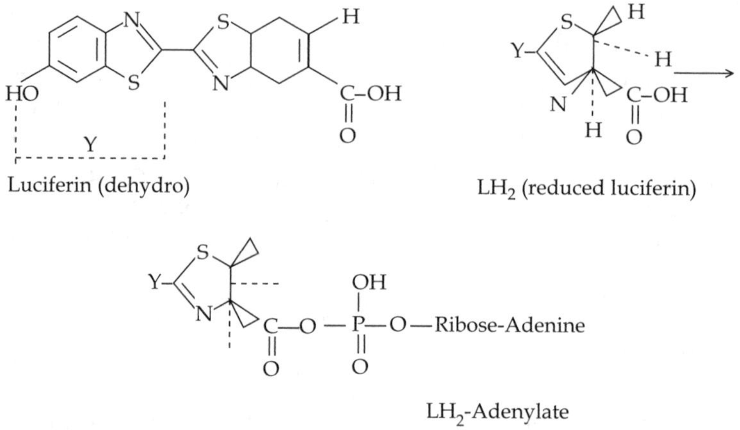

Luciferin (dehydro)

LH_2 (reduced luciferin)

LH_2-Adenylate

Figure 1.13   Firefly bioluminescence.

1.16   Hydrogen

The simplest of the elements, containing a single proton and electron each, of mass almost unity, is the first member of the periodic table. Data may vary from different sources; solid at 4.2 K (d 0.089), H has the atomic number (AN) 1, atomic weight (AW) 1.008 g, melting point (mp) $-259.14°C$, and boiling point (bp) $-252.87°C$ (d 0.071 at 20.4 K). He has a AN 2, AW 4.0026 g, mp $-272.2°C$ (20 atm), and bp $-268.93°C$ (specific gravity 0.124).

Commercial consumption at present is mostly in synthetic fuels, say from coal, mineral oils, petroleum reformation (refineries), and iron and copper ore reductions. Hydrogen is very important because of the versatility of its physical, chemical, and biological properties. More importantly for our purposes, is its potential as a source of energy. Hydrogen liquefies at 33.2 K, 12.8 atm, and 0.03 g/mL and occupies a negligible volume (22.4 times less), compared to its gaseous state. Solid hydrogen and helium are academic ideas. When hydrogen combines with oxygen in a volume ratio of 2:1, heat is generated and the product is water in a vapor state. The reaction in a vapor state occurs with a reduction of volume to $1/3$ and water vapor to water $1/22$, which means the reaction is favored at a higher pressure; alternatively, the change in volume is compensated by utilizing some of the heat that evolves. The calculations are already there. Hydrogen as a combustion fuel or as a material for a fuel cell is less attractive than the fusion reaction such as that which occurs in the sun. Taking it as a model, we may be able to harness huge amounts of thermal and traditional energies, but we should also learn how to manage and handle such enormous outbursts of energy. Two protons fuse to yield a deuterium, a positron, and a neutrino; the last one is the clue to the release of energy that is not yet fully understood by science;

$$2H^1 \rightarrow e^+ + \nu + H^2 \qquad H^2 + H^1 \rightarrow H^3 + \nu \qquad 2H^3 \rightarrow He^4 + 2H^1$$

Solar constant = 1.968 cal/$(cm^2 \cdot min)$ = 3.86×10^{33} erg/s = 1.373 kW/m^2; even at such a long distance, we are unable to use all the energies.

Hydrogen in absence of air or oxygen, or in vacuum, will not burn, but may have a kind of combustion to produce ammonia in air or nitrogen. Combustion of hydrogen in our atmosphere does not produce simple water vapor, but mixture of others, i.e., ammonia and NO_{xS} (nitrogen and oxygen combine at the vicinity of high temperature generated).

Cryogenic and space research have taught us many more lessons. Liquid hydrogen can be stored in special containers (cylinders), or transported through pipes, and is almost an ideal fuel for rockets and spaceships, perhaps next to azides. But at higher altitudes or in space, in the absence of atmosphere, optimal liquid oxygen is also needed to perform the dynamism or thrust. Water vapor is transformed into ice particles instantly due to the very low temperature in space. Liquid hydrogen for such research or experiment is generated at a very high cost, i.e., electrolytic splitting of water. The alternate resource of hydrogen is a by-product in the caustic soda plant. A similar minor and indirect source of hydrogen is water gas ($C + H_2O \rightarrow CO + H_2$), almost obsolete for any large-scale production. None of these examples are renewable in nature, continue to be energy and labor intensive, and cannot stand as competitors as fuel or energy resources. Other commercial sources of hydrogen are dependent on the existing limited supply of natural resources, i.e., coal, naphtha, and natural gas, which are not renewable. The materials are mainly based on fluidization or gasification of coal, and reformation by superheated steam or from steam–iron process ($3Fe + 4H_2O \rightarrow F_3O_4 + 4H_2$); these processes can be broadly classified into (a) thermochemical or solar gasification and (b) fast pyrolysis or other novel gasification. These processes may be totally or partly catalytic. The basic chemical principles are mostly similar to those of classical water gas: $C + H_2O \rightarrow CO + H_2$; $CO + H_2O \rightarrow CO_2 + H_2$. Major sources of hydrogen at present are directly or indirectly natural gas; electrolysis; pyrolytic, thermal, and superheated steam; or geothermal, solar, ocean current, ocean thermal gradient, and nuclear reactors. Biomass as a source of hydrogen as well as energy has been discussed in Sec. 1.2.

1.16.1 Microbial conversion

Many or most organic cellulosic matter, after proper mechanical treatment (homogenizing), can be put to microbial conversion for (a) biomethanation and/or (b) hydrogen production.

1. Biomethanation can utilize human or animal excreta as well as mixed green/organic wastes. This part has been discussed earlier in Secs. 1.12 and 1.13.

2. Hydrogen production is discussed hereafter.

Biohydrogen.   Major routes are

1. Enzymatic (partly microbial) through microbial routes
2. *Klebsiella* and *Clostridium* groups of microbes
3. Different cyanobacteria (blue-green algae)
4. Various photosynthetic bacteria
5. Many aerobes, i.e., bacilli and alkaligenes
6. Facultative groups, i.e., enterobacters, and coli forms
7. Various anaerobes, i.e., rumens, methanogenic, methylotropes, and clostridia

Enzymatic. Glucose dehydrogenase oxidizes glucose into gluconic acid and NADPH, which helps the reduction of H^+ by hydrogenase. Glucose dehydrogenase and hydrogenase are purified from *Thermoplasma acidophylium* and *Pyrococcus furiosus* (optimal growth at 59°C and 100°C, respectively) (Woodward).

Based on metabolic patterns, the microbial systems may be of four types:

1. Photosynthetic microbes evolving H_2 mediated through NADPH (Nicotine Adenine Dinucleotide Phosphate [Coenzyme II-reduced]) by photoenergy.

2. Cytochrome systems operating in facultative anaerobes that convert mainly formates to H_2.

3. Cytochrome containing strict anaerobe, *Desulfovibrio desulfuricans*.

4. Clostridia, micrococci, methanobacteria, and others, without cytochrome, anaerobic heterotrophs.

Klebsiella oxytocae. ATCC (American Type Culture Collection) 13182 can convert formates to H_2 (100%), but only 2 moles of H_2 for each mole of glucose (5%). *C. butyricum* can convert glycerol to 1,3-propanediol, butyric acid, 2,3-butanediol, formic acid, and CO_2 and H_2. *Klebsiella pneumoniae* can convert glycerol into 1,3-propanediol, acetic acid, formic acid, and CO_2 and H_2. The presence of acetate enhances the production of butyrates and H_2, and less propanediol.

Before discussing cyanobacteria and photosynthetic bacteria, we should review the basic reactions involved in photosynthesis, i.e., steps in so-called photophosporylation:

$$H_2O + NADP^+ + PO_4 + ADP \xrightarrow{+h\upsilon} O_2 + NADPH + H^+ + ATP$$

$$CO_2 + NADPH + H^+ + ATP \xrightarrow{-h\upsilon} (HCOH)_n + NADP^+ + ADP + PO_4$$

$$\text{Aerobic: } 6CO_2 + 6H_2O \xrightarrow{+h\upsilon} C_6H_{12}O_6 + 6O_2$$

$$\text{Anaerobic: Isopropanol or } H_2S + CO_2 \xrightarrow{+h\upsilon} \text{Acetone or S } + (CH_2O)_n \, H_2O$$

Cyanobacteria. Popularly known as blue-green algae, and justifiably so (they consume CO_2 and evolve O_2), they are bacteria (absence of nuclei, mitochondria, chloroplasts, etc.) as well as algae.

Cyanobacteria are oxygenic photoautotrophs, possessing photo I and II systems. Cyanobacteria have been well studied, and the details of their physiology and biochemistry are available in reviews and books. They are held by many scientists as potential sources of chemicals, biochemicals, food, feed, and fuel. Most of them are molecular nitrogen fixers and possess a nitrogenase system for H_2 production. They are found to be symbiotic to cycads, lichens, and so forth. Some are hetero-cystous, lacking photolysis of water, and produce H_2 through the nitro-genase step (when N_2 is low). The nonheterocystous species produce H_2 at higher efficiency at low N_2 and O_2 concentrations. Some of the species favor anoxic and dark conditions, but with the presence of organic sub-strates. They may even use sulfides as a source of electrons under an anaerobic environment. They are highly adaptable to a changing envi-ronment and are widely found in salty or sweet water, deserts, hot springs (up to 75°C), as well as Antarctica. Some heterocystous *Anabaena* exhibit H_2 production in an atmosphere of argon and absence of molecular nitrogen. This was the clue to the knowledge that the enzyme nitrogenase, the main biocatalyst for molecular nitrogen fixa-tion, is present in cyanobacteria and is the key route of H_2 production:

$$N_2 + 8H^+ + 8e^- + 12ATP \rightarrow 2NH_3 + H_2 + 12ADP + 12Pi$$

A "reversible hydrogenase" (in photolysis of water, $2H_2O \rightarrow 2H_2 + O_2$), is present in both heterocyst and vegetative cells and produces H_2 at a lower rate than a nitrogenase. An "uptake hydrogenase" also operates (minor) connected to cytochrome chain, providing both H^+ and elec-trons. H_2 evolution is common, but the photolytic O_2 is inhibitory to nitrogenases, which is protected by other biochemical and structural alternatives existing in heterocysts.

Large amounts of ATP, which is required for the reaction are gener-ated in the event of photosynthesis and respiration. The electron (reduc-tant) supply in the nitrogenase equation comes from metabolites, i.e., amino acids, mainly from carbohydrates (maltose, glucose, fructose, other pentoses, tetroses, etc.), produced and stored in the vegetative cells through photo I and II systems.

Nitrogenase Co II, i.e., NADPH (gained through the pentose phosphate route) happens to be an electron donor through NADP oxidoreductase/ferredoxin or flovodoxin. Other electron-supplying batteries are also envisaged.

1. Through uptake hydrogenase–ferredoxin (photoactivated)

2. Through pyruvate–ferredoxin oxidoreductase

3. Reduced ferredoxin from isocitrate dehydrogenase

4. NADH generated in the glycolytic route

Under anaerobic or low aerobic conditions, nitrogenase activity may exist in vegetative cells, but H_2 generation is of poor order.

Photosynthetic bacteria.   Hydrogen production is guided by the surplus of ATP and reductant organic metabolites (carbon sources from the

Krebs cycle) and reduced nitrogen sources (glutamate/aspartate). Interactions of hydrogenase and nitrogenase may be complementary or competitive in different species or mutants. Nitrogenase (Mo, Ni, or Fe) also with mixed isozymes are reported. Some mutants liberate H_2 more efficiently, utilizing DL-malate, D-malate, and L-lactate. Photoautotrophic growth is found to be less efficient in producing H_2 than photoheterotrophic growth with limited nitrogen in nutrients. Normally, in photosynthetic bacteria, hydrogenase utilizes the hydrogen as a reductant for CO_2 fixation and also for fixing molecular nitrogen. Nitrogenase reduces molecular nitrogen, along with the production of molecular hydrogen at the expense of almost six stoichiometric equivalents of ATP. This means that concurrent nitrogenase activity during photosynthesis competitively consumes the ATP that is produced and lowers the CO_2-fixing efficiency.

Rhodospirillum and *Rhodopseudomonas* grow aerobically in the dark. But *Rhodospirillum rubrum* growing on glutamate (a nitrogen source) exhibit good hydrogen release during photosynthesis. Quantitative production of hydrogen has also been observed, growing on acetate, succinate, fumarate, and malate, by photosynthesis, initially in the presence of limited ammonium salts.

In *Rhodopsuedomonas acidophilla*, hydrogenase and nitrogenase are genetically linked. Several species of Rhodospirillaceae can perform nonnitrogenase-mediated hydrogen production in the absence of light, using glucose and organic acids including formates. Different strains of *Rhodopseudomonas gelatinus* and *Rhodobacter sphaerolides* exhibit highly efficient production of hydrogen [90 μL/(h $\cdot$ mg) cell] grown in a glutamate–malate medium.

In some cultures of *Rhodopseudomonas capsulata, R. rubrum,* and *Rhodomicrobium vannielli,* replacement of glutamate by N_2 gas improved productivity of H_2 (760 mL/d, 10 days) decreasing a little on aging. The model of a nozzle loop bioreactor, with immobilized *R. rubrum* KS–301 in calcium alginate, initial glucose concentration of 5.4 g/L, 70 h at 30°C, showed production of hydrogen 91 mL/h (dilution rate of 0.4 mL/h). Improvement was suggested by using an agar gel for immobilization.

Aerobes.

1. *Bacillus licheniformis* isolated from cattle dung showed production of H_2 in mixed culture media. Immobilized on brick dust, the aerobe maintained H_2 production for about 2 months in a continuous system, with an average bioconversion ratio of 1.5 mole of H_2 per mol of glucose.

2. *Alcaligenes eutrophus*, when grown on gluconates or fructose anaerobically, produces H_2. Hydrogenase directly reduces the coenzyme

using hydrogen, and the excess hydrogen is spilled out. Higher concentration of formate reduced hydrogen production.

$$HCOOH \leftrightarrow CO_2 + H_2$$

Facultative anaerobes.

1. *Enterobacter: Enterobacter aerogenes*, as an example, can use varied and mixed nutrients, i.e., glucose, fructose, galactose, mannose, peptones, and salts (pH 4.0, 40°C); and may show activity for about a month in a continuous culture; evolution of hydrogen was about 120 mL/h/L of medium; 0.8 mol/mol of glucose. Accumulation of acetic, lactic, or succinic acids is likely to cause antimetabolic suppression in older cultures.

2. *Escherichia coli:* Anaerobically, it can use formate to produce CO_2 and H_2. Carbohydrates as nutrient sources usually end up with mixed products, i.e., ethanol, acetate, hydrogen, formate, carbon dioxide and succinate.

Various anaerobes.

1. *Ruminococcus albus* mostly converts cellulose to CO_2, H_2, HCOOH, C_2H_5OH, CH_3, and COOH. Pyruvatelyase may be functional in the production of H_2 (237 mol/mol of glucose). Further details are not available.

2. *P. furiosus* (thermophilic archeon) possesses nickel-containing hydrogenase and produces hydrogen using carbohydrate and peptone, at 100°C. The metabolic system seems to be uncommon to those of nonthermophiles.

3. *Methanobacterium (Methanotrix) soehngenii* (methanogens) can grow on acetate and salts media, but can split formate into hydrogen and carbon dioxide. *M. barkeri*, in the presence of bromoethane sulphonate, has suppressed methane production; instead, hydrogen, carbon dioxide, carbon monoxide, and water were produced.

4. *Methylomonas albus* BG8 and *Methylosinus trichosporium* OB3b (methylotrophs) used various substrates, i.e., methane, methanol, formaldehyde, formate, pyruvate, and so forth. But formate was found to be most useful for production of hydrogen under anaerobic conditions.

5. *C. butyricum, C. welchii, C. pasturianum, C. beljerinscki,* and so forth are very efficient in utilizing different carbohydrate sources and even effluents to produce hydrogen (see Fig. 1.14). Immobilization of these cells has also been successful.

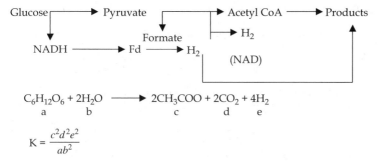

$$C_6H_{12}O_6 + 2H_2O \longrightarrow 2CH_3COO + 2CO_2 + 4H_2$$
$$\quad a \qquad b \qquad\qquad\quad c \qquad\quad d \qquad e$$

$$K = \frac{c^2 d^2 e^2}{ab^2}$$

Figure 1.14 Hydrogen production.

Removal of or reducing concentrations of either CO_2 or H_2 or the combination of both is likely to favor a forward reaction, i.e., to improve production of H_2. Attempts to remove CO_2 by collecting the evolved gases through 25% (w) NaOH solution, using *E. aerogenes* (E 82005), showed better production of H_2, which improved further by enriching nitrogenous nutrients in the culture media—from 0.52 moles of H_2 per mole of glucose, increased to 1.58 moles [9].

Similar attempts are made using *E. cloacae* and reducing the partial pressure of H_2 during the production of gases, by reducing the operating pressure of the reactor and simultaneous removal of CO_2 [through 30% (w/v) KOH], maintaining an anoxic condition by flushing Ar at the onset [10]; by reducing the operating pressure to 0.5 atm, the molar ratio of H_2 yield per mole of substrate doubled (1.9–3.9). Other technical and economic benefits were also cited. There are other similar claims of improved biohydrogen production [11], using altered nutrients (20 g of glucose, 5 g of yeast extract, and 5 g/L of tryptone) and different mutants of *E. aerogenes* HU-101. HU-101 and mutants A1, HZ3, and AAY, respectively yielded 52.5, 78, 80, and 101.5 mmol of hydrogen per liter of media.

References

1. R. Buvel, M. J. Allen, and J. P. Massue. *Living Systems as Energy Converters*, Amsterdam: North Holland, 1977.
2. H. Baltscheffsky. *Origin and Evolution of Biological Energy*, Amazon Company: United Kingdom, 1996.
3. L. A. Kristofferson and V. Bokalders. *Renewable Energy Technologies: Their Applications in Developing Countries*. A Study of the Beijer Institute and the Royal Swedish Academy of Sciences, Oxford, Pergamon: Sweden, 1986.
4. M. Calvin. Photosynthesis as a resource for energy and materials, *American Scientist* **64**, 270–278, 1976.
5. A. Nag and K. Vizaykumar. *Environmental Education and Solid Waste Management*, New Age Publisher: New Delhi, India, 2006.
6. *Proceedings of Bio-Energy Society* (Department of Nonconventional Energy), CGO, New Delhi, India, October 14–16, 1985.

7. J. G. Zeikus, Biology of methanogenic bacteria, *Bacteriological Reviews* **41**, 514–541, June 1977.

8. R. C. Kuhad and A. Singh. Lignocellulose biotechnology: Current and future prospects, *Critical Reviews in Biotechnology* **13**, 151–172, 1993.

9. S. Tanisho, M. Kuromoto, and N. Kadokura. Effect of CO_2 removal on hydrogen production by fermentation, *International Journal of Hydrogen Energy*, **23**(7), 559–563, 1998.

10. D. Das, B. Mandal, and K. Nath. Improvement of biohydrogen production under decreased partial pressure of H_2 by *Enterobacter cloacae*; *Biotechnology Letters* **28**, 831–835, 2006.

11. M. Abdul Rahaman, Y. Furutani, Y. Nakashimada, T. Kakizono, and N. Nishio. Enhanced hydrogen production in altered mixed acid fermentation of glucose by *Enterobacter aerogenes*, *Journal of Farm and Bioengineering* **41**(4), 356–363, 1997.

Photosynthetic Plants as Renewable Energy Sources

Ahindra Nag and P. Manchikanti

2.1 Introduction

Renewable energy is an energy resource naturally regenerated over a short time scale derived from the sun (such as thermal, photochemical, and photoelectric) or from other natural environment effects (geothermal and tidal energy). It is forecasted that approximately half of the total resources in the world will be exhausted by 2025. This survey has also revealed that global warming and climate change are serious issues that need immediate action. The use of fossil fuels (coal, oil, gas, etc.) contributes significantly to global warming and climate change [1]. Worldwide there is strong support for renewable energy, as proven by a number of surveys [1, 2]. In 2003, a European Commission survey across the 15 European Union (EU) countries showed that 69% of the citizens supported more renewable energy-related research, compared to 13% for gas, 10% for nuclear fission, 6% for oil, and 5% for coal. Understandably, due to the inherent recycling nature as well as environmental benefits involved, renewable sources of energy are the solution for energy management. There is an increased investment globally in such technologies for not only enhancing the preservation of biological resources but also for increasing energy efficiency and pollution control [1].

Biomass is one such renewable source of energy. Out of the 1.1×10^{20} kW heat generated every second by the sun, only 47% (~7×10^{17} kWh) reaches the earth's surface. Solar energy is utilized by conversion to different energy forms such as biomass, wind, or hydropower. Green plants are only able to effectively use visible light of wavelength falling between

400 and 700 nm. This photosynthetically active radiation constitutes about 43% of the total incident solar radiation to produce biomass. Biomass energy generally involves the utilization of energy contents of such items as agricultural residues (pulp derived from sugarcane, corn fiber, rice straw and hulls, and paper trash) and energy crops. So, biomass is a comprehensive term that includes essential forms of matter derived from photosynthesis or ultimately available as animal waste [2]. The production of energy from plants is not a new idea; wood burning has been in common use since ancient times. About one-seventh of the energy used around the world is derived from firewood. Biomass supplies 14% of the world's primary energy consumption and is considered to be one of the important renewable resources of the future. With the increase in population and the demand for resources, demand for biomass is expected to increase rapidly. On average, 38% of the primary energy resources in developing countries is biomass. In the United States alone, biomass sources provide about 3% of all the energy consumed. In terms of energy efficiency measures and stabilization of energy consumption between 2010 and 2020, the European Renewable Energy Council (EREC) survey estimates that among the various types of renewable energy resources, biomass-derived energy will be a significant portion of energy used [1]. The survey also reveales that biomass and biofuels are the top two in terms of employment that they generate. Burning new biomass does not contribute to new CO_2 into the atmosphere as replanting harvested biomass ensures that CO_2 is absorbed and returned for a cycle of new growth [2].

2.2 Mechanism and Efficiency of Photosynthesis in Plants

In photosynthesis, CO_2 from the atmosphere and water from the earth combine to produce carbohydrates, which are the components of biomass and solar energy that drive this process. When biomass is efficiently utilized, the oxygen from the atmosphere combines with the carbon in plants to produce CO_2 and water (see Fig. 2.1). Typically, photosynthesis converts less than 1% of the available sunlight to be stored as chemical energy.

The advantages of using plants for renewable energies (fuels and chemicals) are listed follows:

- Advances in agriculture and forestry technologies have resulted in increased utilization of land resources for cultivation of energy crops.
- By increasing harvesting of solar energy, there is effective usage of biomass-based resources.

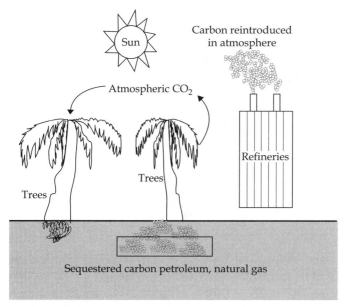

Figure 2.1 Simplified carbon cycle.

- Multiple economic benefits can be derived—for example, sugar can be used as such for fermentation to alcohol—depending on the market.
- Biomass combustion, unlike fossil fuels, does not contribute to increased CO_2 levels in the atmosphere [2].
- Increased employment opportunities resulting from the above.

While the advantages of using biomass-based energies are apparent, it is important to note that biomass cannot by itself provide complete replacement of fossil fuels. Hence, it is one of the solutions toward achieving energy efficiency. Further factors, such as competition for biomass between energy production and human nutritional needs, as well as the possible environmental effects, must be kept in mind. There are several factors that should be considered in using plants for the generation of energy; efficiency of solar energy absorption and conversion, quality of biomass produced, plant growth, growth under marginal conditions, soil characteristics, and cost-effectiveness of production of energy and conversion. We will focus on the utilization of terrestrial plants for production of renewable energies.

2.3 Photosynthetic Process

There are essentially two types of reactions in photosynthesis: a series of light-dependent reactions that are temperature independent (or light reaction) and a series of temperature-dependent reactions that are light

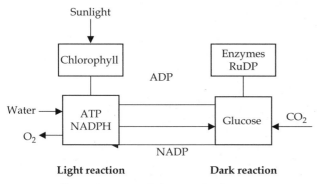

Figure 2.2 General process of photosynthesis.

independent (or dark reactions). The rate of the light reaction can be increased by increasing light intensity, and the rate of the dark reaction can be increased by increasing temperature to a certain extent (see Fig. 2.2).

2.3.1 Hill reaction (light reaction)

The process of formation of CO_2 and O_2 during photosynthesis is called the Hill reaction or photolysis of water. This primary photochemical reaction takes place in the presence of sunlight. The reaction is associated with chlorophyll, and after receiving light energy, the chlorophyll becomes activated. The steps in the Hill reaction can be summed up in the following manner:

1. **Absorption of light and activation of chlorophyll** Radiant light contains very tiny energized particles called photons or quanta, which are absorbed by the chlorophyll and it becomes activated.

2. **Photolysis** Is the dissociation of water molecules by light energy that have been absorbed by the chlorophyll. The reaction can be represented as

$$4H_2O \xrightarrow{\text{Light energy}} 4H^+ + 4OH^-$$

$$4H^+ \xrightarrow{\text{Chlorophyll}} 2H_2$$

$$H_2 + NADP \rightarrow NADPH_2 \text{ (Hydrogen acceptor)}$$

$$4OH^- \xrightarrow{\text{Recombination}} 2H_2O + O_2$$

3. **Photophosphorylation**   This is the stage of formation of ATP from ADP

$$ADP + Pi \rightarrow ATP$$

2.3.2 Blackman's reaction (dark reaction)

The dark reaction is independent of light. This reaction is purely enzymatic and is carried out in the stoma portion of the chloroplast. Ribulose-1, 5-diphosphate (RuDP), a pentose phosphate present in plant cells, acts as the initial acceptor of CO_2 and changes thereby into a very unstable C_6. The latter is converted into 3-phosphoglyceric acid (3-PGA), which is transferred to 3-phosphoglyceraldehyde. For this reaction, ATP and $NADPH_2$ (produced in the light reaction) are necessary as cofactors. Three molecules of RuDP combine with three molecules of CO_2 to give rise to six molecules of PGA. Three molecules of RuDP utilized initially as CO_2 acceptors are regenerated by five molecules of phosphoglyceraldehyde through different intermediates like xylulose-5-phosphate and ribulose-5-phosphate. The only molecule of phosphoglyceraldehyde is converted into fructose-1,6-diphosphate, which may be transformed into sucrose and starch through other reactions.

2.3.3 Efficiency of photosynthesis

While there are several factors that affect photosynthetic rate, the three main factors are light intensity, carbon dioxide level, and temperature. The net efficiency of photosynthesis is estimated by the net growth of biosynthesis and the amount used for respiration. The requirements for achieving high energy conversion are optimal temperature, light, nutrition, leaf canopy, absence of photorespiration, and so forth. Many plant species can be distinguished by the type of photosynthetic pathway they utilize. Most plants utilize the C_3 photosynthesis route. C_3 determines the mass of carbon present in the plant material. Poplar, willow, wheat, and most cereals are C_3 plant species. Plants such as perennial grass, *Miscanthus*, sweet sorghum, maize, and artichoke all use the C_4 route of photosynthesis and accumulate significantly greater dry mass of carbon than the C_3 plants. Advances in crop production, agricultural techniques, and so forth have led to potential applications in low-cost biomass production with high conversion efficiencies. Further, introduction of alternative nonfood crops on surplus land and the use of biomass as a sustainable and environmentally safe alternative make biomass an attractive renewable energy resource. The potential of biomass energy derived from forest and agricultural residues worldwide is estimated at about 30 EJ/yr. For the adoption of biomass as a renewable energy source, the cultivation of energy crops using fallow and marginal land and efficient processing methods are vital [3].

C_3 metabolism in plants and the pentose phosphate pathway. In C_3 plants, the pathway for reduction of carbon dioxide to sugar involves the reductive pentose phosphate cycle. This involves addition of CO_2 to the pentose bisphosphate, ribulose-1,5-bisphosphate (RuBP). The enzyme-bound carboxylation product is hydrolytically split, through an internal oxidation-reduction process, into two identical molecules of 3-PGA. An acyl phosphate of this acid is formed by reaction with ATP. This is further reduced with NADPH. Five molecules of the resulting triose phosphate are converted into three molecules of the pentose phosphate, ribulose 5-phosphate. Three molecules of ribulose 5-phosphate are converted with ATP to give the carbon dioxide acceptor, RuBP, thereby completing the cycle. When these three RuBP molecules are carboxylated and split into six PGA molecules and these are reduced to triose phosphate, there is a net gain of one triose phosphate molecule over the five needed to regenerate the carbon dioxide acceptor. Triose phosphate is formed in this cycle and can either be converted into starch for storage of energy inside the chloroplast, or it can serve its primary function by being transported out of the chloroplast for subsequent biosynthetic reactions. In a mature leaf, sucrose is synthesized and exported to the rest of the plant, thus providing energy and reduced carbon for growth [4]. Wheat, potato, rice, and barley are examples of C_3 plants. A representative C_3 cycle is shown in Fig. 2.3.

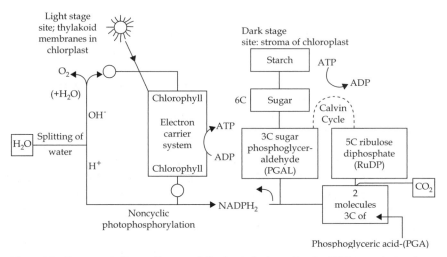

Figure 2.3 Representation pathways of C_3 plant photosynthesis. (*With permission from Oxford University Press.*)

C$_4$ metabolism in plants.   In air that contains low carbon dioxide in relation to oxygen, oxygen competes for the carbon dioxide binding site of the ribulose bisphosphate carboxylase. This is known to set off a process of photorespiration in plants, and it is believed that the C$_4$ plants have evolved from such a mechanism. Such plants possess a specialized leaf morphology called "Krantz anatomy" and a special additional CO$_2$ transport mechanism. This typically overcomes the problem of photorespiration. Such avoidance of photorespiration is known to result in higher growth rates. The Krantz anatomy is characterized by the fact that the vascular system of the leaves is surrounded by a vascular bundle, or bundle-sheath cells, which contain enzymes of the reductive pentose phosphate cycle. The reduction of CO$_2$ is similar to that of C$_3$ plants, except that the CO$_2$ for carboxylation of CO$_2$ is derived not from the stomata but is released in bundle-sheath cells by decarboxylation of a four-carbon acid (C$_4$ acid). This C$_4$ acid is supplied by the mesophyll cells that surround the bundle sheath cells. The C$_4$ pathway for the transport of CO$_2$ starts in a mesophyll cell with the condensation of CO$_2$ and phosphoenolpyruvate to form oxaloacetate, in a reaction catalyzed by phosphoenolpyruvate carboxylase (PEPCase), and the reduction of oxaloacetate to malate [5]. Figure 2.4 shows the C$_4$ cycle of CO$_2$ fixation in photosynthesis.

Due to the elimination of the photorespiration process, C$_4$ plants are proposed to be ideal for increased biomass production especially in marginal conditions. Grasses are suitable for this purpose as they can be

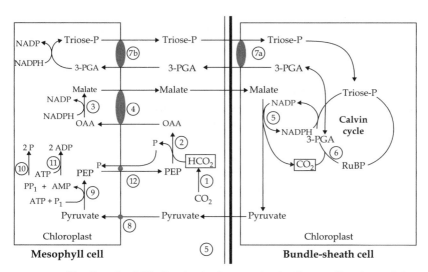

Figure 2.4   The C$_4$ cycle of CO$_2$ fixation in photosynthesis. (*Source: Häusler et al. [5]*)

TABLE 2.1   Differences between C_3 and C_4 Plants

Plant characteristics	C_3 cycle type	C_4 cycle type
Leaf anatomy	Mesophyll (palisade and spongy type), no chloroplasts in bundle-sheath cell	Krantz anatomy, bundle-sheath cell with chloroplasts
Chloroplasts	Single-type	Dimorphic
Carboxylase type	Primary (Rubisco)	Primary PEPCase in mesophyll, Secondary (Rubisco in bundle-sheath cell)
Primary CO_2 acceptor	RuBP	PEP
Primary stable product	3-phosphoglyceric acid (3-PGA)	Oxalocetate (OAA)
Ratio of CO_2:ATP:NADPH	1:3:2	1:5:2
Productivity (ton/ha · yr)	~20	~30

grown on a repetitive cropping mode for continuous and maximum production of biomass. Grasses such as Bermuda grass, Sudan grass, sugarcane, and sorghum are good candidates for energy generation from biomass. A comparison of the characteristics of C_3 and C_4 plants, in terms of leaf anatomy, is shown in Table 2.1.

2.4   Plant Types and Growing Cycles

Several plants have been proposed to be good sources of energy. These include woody crops and grasses/herbaceous plants, starch and sugar crops and oilseeds, fast growing trees such as hybrid poplars, shrubs such as willows, and so forth. Energy crops can be grown on agricultural lands not utilized for food, feed, and fiber. Farmers could plant these crops along the riverbanks, along lakeshores, between farms and natural forests, or on wetlands. These crops could be a good source of alternate income, reducing the risk of fluctuating markets and stabilizing farm income. Woody plants, herbaceous plants/grasses, and aquatic plants are different sources for biomass production. The type of biomass selected determines the form of energy conversion process. For instance, sugarcane has high moisture content, and therefore, a "wet/aqueous" bioconversion process, such as fermentation, is the predominant method of use. For a low-moisture content type such as wood, gasification, pyrolysis, or combustion are the more cost-effective ways of conversion.

Characteristics of an ideal energy crop are mentioned below:

- Low energy input to produce
- Low nutrient requirements
- Tolerance to abiotic and biotic stresses

- High yield/high conversion efficiency
- Low level of contaminants

Energy plantations and cropping are means of growing selected species of trees or crops that can be harvested in a shorter time for fuel, energy, and other resources. Each type of popular plant species is discussed in brief, with respect to renewable resources.

Euclayptus. It is a fast growing plant for firewood (see Fig. 2.5). Different species such as *Eucalyptus nitens, E. fastigata,* and *E. globulus* are used in many countries such as Australia and Brazil. Eucalyptus, an exotic species from Australia, is a versatile tree which adopts itself to a variety of edaphic and climatic conditions. It comes up in different types of soils and climates varying from tropical to warm temperatures and with annual rainfall ranging from 400 to 4000 mm. It grows well in deep, fertile, and well-drained loamy soils with adequate moisture. A large eucalyptus plantation program has been successfully launched in Brazil to serve as the feedstock for its methanol plant. Amatayakul et al. suggest that if eucalyptus wood is used for electricity generation, the cost of electricity generation would be 6.2 US cents/kWh, and consequently, the cost of substituting a wood-fired plant for a coal-fired plant and a gas-fired plant would be US $107 and $196 per ton of C, respectively [6]. Eucalyptus plantations could offer economically attractive options for electricity generation and CO_2 abatement.

Casuarina. *Casuarina* is a genus of shrubs and trees of the Casurinacea family, native to Australia and islands of the Pacific. The species involve *Casuarina equisetifolia* Linn. It is a big evergreen tree with a trunk diameter of 30 cm and height 15 m, and is harvested after 5–7 years (see Fig. 2.6). The plant fixes nitrogen through symbiotic bacteria and thus adds fertility to the soil. It is very useful for afforesting sandy beaches and sand dunes. The wood is used for fuel purposes.

Figure 2.5 Eucalyptus plantation.

Figure 2.6 *Casuarina* plantation.

Mimosa.   *Mimosa leucocephala* or kubabul is a fast-growing species known for energy plantation (see Fig. 2.7). It has a very high potential for nitrogen fixation and can be well adapted to poor soils, drought, and windstorms. It can fix up to 500 kg of nitrogen per hectare per annum. It coppices readily, and the sprouts, after harvesting, can grow up to 18 ft in just 1 year. It is also called the wonder tree. Under irrigated conditions, it can give fodder yields up to 80–100 ton/(ha · yr). Three different

Figure 2.7 Mimosa plantation. (*Source: Creative Commons.*)

varieties of this species (Hawaiian, Cunningham, and Brazilian) are commonly used for plantations in Hawaiian, Salvador, and Peru. The Hawaiian and the Cunningham varieties are used for energy plantation in India and Australia, respectively. A Hawaiian plantation of 1.27 hectares can support a 1-MW power plant. In Brazil and the Philippines, it is converted into charcoal that has 70% of the heating value of oil. Charcoal can be used to produce calcium carbide, acetylene, vinyl plastics, pig iron, and ferroalloys. The low silica, ash, and lignin contents and high cellulose content make this plant good for paper and pulp materials, and also for rayons and cellophanes. It not only gives a prolific fuelwood yield but is also a nutrient-rich fodder for livestock.

Sugarcane.   Sugarcane (*Saccharum officinale*) is a hardy plant that can tolerate poor drainage, can be cultivated as a rotation crop, and can be maintained for years. It is grown in fertile areas with more than 1000 mm of rain and an abundant supply of water. The ethanol yields from this are in the range of 3.8–12 kL/(ha · yr) [7].

Cassava.   Cassava (*Manihot esculenta*), like sugarcane, is grown in tropical climates with an average rainfall of 1000 mm. As it is relatively drought resistant, it can withstand lower annual rainfall. It needs to be grown annually and is difficult to mechanize, and compared to sugarcane, it is less energy efficient. Ethanol yields are estimated in the range of 0.5–4.0 kL/(ha · yr).

Sorghum.   *Sorghum* embraces a wide variety of plant types and, unlike sugarcane and cassava, is found in the tropical summer rainfall zones. While it can grow in as little as 200–250 mm annual rainfall, maximum yields are obtained in a minimum of 500–600 mm rainfall. Compared to other cereals, it can tolerate high temperatures. Due to its deep root system and low rate of transpiration, it is exceptionally resistant to drought. Ethanol yields of stems and grains of sorghum are in the range of 1.0–5.0 and 2.0–5.0 kL/(ha · yr), respectively.

Babassu.   Babassu (*Orbignya* sp.) is a palm popular in Brazil for the ethanol derived from it. The mesocarp of coconut is the raw material for ethanol production, with an estimate of 0.24 kL/(ha · yr).

Oil-bearing crops.   Vegetable oils are the most promising alternatives to diesel fuel. About 97% of all oil-bearing plants are grown in tropical and subtropical climates. There has been some research into the use of plant oils from sunflower, peanut, rapeseed, soybean, and coconut oils as biofuels in unmodified/slightly modified engines. Seed-based oils are shown to lead to slightly higher fuel consumption, probably due to their calorific value [8]. About 14% of the oil supplied in the world market is palm oil, yielding an average 3.4 ton/(ha · yr) of oil [9]. Individual palm seeds, however, are capable of producing much higher yields. The extraction of palm kernel oil increases fuel oil yields by 10%. Current cultivation is mostly in lowland humid tropics such as Malaysia, West Africa, and Indonesia. While the conditions to grow coconut palms are similar to oil palms, the yield potential of coconut palms has not yet been developed to that potential. Soybeans and peanuts are annual leguminous crops that are used as sources of both oil and protein. Soybeans thrive best in subtropical climates. The individual varieties differ greatly in terms of their reaction to the length of a day and normally can be grown in a limited geographical area. Peanut cultivation requires an ambient temperature for growth, as less than optimal temperatures are known to result in poor yields. Due to its deep root system, it is relatively resistant to drought. It is also a suitable crop for mixed cultivation along with oil palms and corn. In terms of calorific value of seed, oil plants such as *Simmondsia chinensis, Pittosporium resinifreum, Ricinus communis, Jatropha curcas,* and *Cucurbita foetidissima* are found to be ideal. Buffalo gourd (*Cucurbita foetidissima*), a desert-adapted plant, produces high-quality oil and fermented starch. The oil has a high ratio of unsaturated to saturated fatty acids. Crude protein and fat content in the whole seeds is 32.9% and 33%, respectively [8]. With a seed yield of 3000 kg/ha and estimated 16% hydrocarbon, about 35 barrels of crude oil could be produced per hectare, in addition to carbohydrate from roots, forage from vines, and protein-rich oil cakes. Jojoba (*Simmondsia chinensis*) is a shrub that grows naturally in the United States and Mexico. Its seeds contain about 50% of oil by weight and does not decrease with long-term storage. The oil is remarkably resistant to degradation by bacteria, probably because it cannot cleave and metabolize the long-chain esters it contains (mostly hydrocarbons containing 38–44 carbon atoms). Jojoba oil has potential uses as a fuel and chemical feedstock, and can also be used as a replacement for vegetable oils in foods, hair oils, and cosmetics since it does not become rancid.

Additionally, it can be used as a source of long-chain alcohols for antifoaming agents and lubricants. The hydrogenated oil is a white, hard crystalline wax and has potential uses in preparation of floor and automobile waxes, waxing fruit, impregnating paper containers, and manufacturing of carbon paper and candles. Physic nut (*Jatropha curcas*), a tropical American species, is a large shrub, or a small tree. The seeds yield 46–58% oil of kernel weight and 30–40% of seed weight. In trade, this oil is called curcas oil. All parts of the plant exude sticky, opalescent, acidic, and astringent latex, containing resinous substances. The bark of this plant is a rich source of tannin (31%) and also yields a dark-blue dye. Now *Jatropha* oil, a semidrying oil, is in high demand for use as biodiesel in Asian countries. It is employed in preparation of soaps and candles and used as an illuminant and lubricant. In China, a varnish is prepared by boiling the oil with iron oxide, and in England, it is used in wool spinning. The oil is used for medicinal purposes for skin diseases, for rheumatism, as an abortifacient, and it is also effective in dropsy, sciatica, and paralysis.

Miscanthus. *Miscanthus,* a thin-stemmed grass, has been identified as an ideal fuel crop as it gives a high dry-matter yield (see Fig. 2.8). Under adequate rainfall conditions, light-arable soils give good yield. It has been found that dark-colored soils produce better yield than light-colored soils. It has been evaluated as a bioenergy crop in Europe for over 10 years and is grown in several European countries. Annual harvesting ability, low mineral content, and good energy yield per hectare are desirable characteristics. It is propagated as rhizomes planted in double rows about 75 cm apart, with 175-cm gaps between the rows. While disease control is not a significant issue, weed control measures are important. In Germany and Denmark, yields are 13–30 ton/ha for 3- to 10-year-old plantation [10].

Panicum. *Panicum virgatum* or switchgrass (see Fig. 2.9) is another thin-stemmed herb that has been used as a model plant [10]. It is a C_4 species, and though it has lower moisture content than wood, it has similar calorific value. It has been found suitable for the development of

Figure 2.8   *Miscanthus. (Source: www.bluestem.ca/ miscanthusgracillimus.htm. Used with permission.)*

Figure 2.9 *Panicum. (Source: www.biology.missouristate.edu/ Herbarium/Plants. Used with permission.)*

ethanol for petrol replacement. The low ash and alkali content makes it a suitable fuel for combustion.

Switchgrass has been identified to be a good model bioenergy species, due to its high yield, high nutrient-use efficiency, and broad geographical distribution. Further, it also has good attributes in terms of soil quality and stability, cover value for wildlife, and low inputs of energy, water, and agrochemicals. Evaluation of the use of switchgrass with coal in existing coal-fired boilers and the handling, operation, combustion, and emission characteristics of the co-firing process have been studied. Switchgrass has supplied up to 10% of the fuel energy input. In comparison to the use of corn for the source of bioethanol, switchgrass has been found to generate 15 times more efficiency of energy production, and it is predicted that switchgrass may entail more profits than conventional crops for a specific area [10].

Hemp. Hemp is a member of the mulberry family that includes mulberry, paper mulberry, and the hop plant (see Fig. 2.10). It has a cellulose content of about 80% and has been grown for the production of medicinal, nutritional, and chemical production. Hemp is the earliest recorded plant cultivated for production of textile fiber. It has a low-moisture content for biomass feedstock [11].

***Artocarpus hirsute* and *Ficus elastica*.** Stem and leaf samples of *A. hirsute* and *F. elastica* have been evaluated for their potential as a renewable energy source. Stem and leaf samples of *F. elastica* and *A. hirsute* were evaluated for polyphenol, oil, and hydrocarbon contents. *F. elastica*

Figure 2.10 Hemp. (*Source: www.greenspirit.com. Used with permission.*)

shows the maximum accumulation of protein (24.5%), polyphenol (4.2%), oil (6.1%), and hydrocarbon (2.0%) contents. The leaf of *F. elastica* has been identified to be a good renewable energy source [12].

Calotropis procera. Latex obtained from *C. procera* could be hydrocracked to obtain hydrocarbons under severe thermochemical conditions. Instead, biodegradation is a less energy-intensive technique for latex degradation. Enhancements in the heptane level have been found in *C. procera* latex that was subjected to different fungal and bacterial treatments, compared to those of untreated ones. Nuclear magnetic resonance (NMR) and fourier transform infrared spectroscopy (FTIR) analyses reveals that the latex has undergone demethylation, dehydrogenation, carboxylation, and aromatization during microbial treatment. Petroleum obtained by hydrotreatment of the biotransformed latex is proposed to be used as fuel [13]. Some of the important latex-bearing plants are *Hevea brasiliensis, Euphorbia* sp., *Parthenium agentatum, Pedilanthus macrocarpus, F. elastica,* and *Manihot glaziorii.* Several resin-rich plants such as *Cappaifera multijuga* (diesel tree), *Copaifera langsdorffi, Pinus, Dipterocarpus, Shorea* sp., and *Pithosporum resiniferum* produce prolific terpene and oleoresins, and are as such very desirable fuel crops. Woody and herbaceous plants have specific growth conditions, depending on the soil type, soil moisture, nutrient content, and sunlight. These factors determine their suitability and growth rates for specific geographical locations. Cereals such as wheat and maize, and perennial grasses such as sugarcane have varied yields with respect to the climatic conditions. Depending on the habitat, plants differ in their characteristic makeup. Their cell walls have varying amounts of cellulose, hemicellulose, lignin, and other minor components. The relative proportion of cellulose and lignin is one of the selection criteria in identifying the suitability of a given plant species as an energy crop. Herbaceous plants are usually perennial, having a lower proportion of lignin that binds together with cellulose fibers. Woody plants characterized by slow growth are composed of tightly bound fibers resulting in their hard external surface. Generally, cellulose is the largest component, representing about 40–50% of the biomass by weight; the hemicellulose portion represents 20–40% of the material by weight. Cellulose is a straight-chain polysaccharide composed of D-glucose units. These units are joined by β-glycosidic linkage between C-1 of one glucose unit and C-4 of the next glucose unit. The number of D-glucose units in cellulose ranges from 300–2500. Hemicellulose is a mixture of polysaccharides, composed almost entirely of sugars—such as glucose, mannose, xylose, and arabinose—and methylglucuronic and galacturonic acids, with an average molecular weight of <30,000 g. Cellulose is crystalline, strong, and resistant to hydrolysis, whereas hemicellulose has a random, amorphous structure with little strength. It is easily hydrolyzed by dilute acid or base.

A complete structure of lignin is not well defined because the lignin structure itself differs between plant species. Generally, lignin consists of a group of amorphous, high-molecular-weight, chemically related compounds. Phenylpropanes, three carbon chains attached to rings of six carbon atoms, are the building blocks of lignin. These might have one or two methoxyl groups attached to the rings. Sugar/starch feedstocks, such as cereals, have been traditionally used in biochemical conversion of biomass to liquids such as ethanol. High-cellulose content of biomass is generally more efficient and therefore preferred over the lignin-rich biomass for conversion of glucose to ethanol. Depending on the end use and type of bioconversion preferred, the choice of the plant species varies. In northern Europe, the C_3 woody species especially grown on short rotation coppice, such as willow and poplar, and forestry residues, are used [14]. In Europe, there is wide interest in the use of oilseed rape for producing biofuel [15]. Brazil was one of the first countries to begin large-scale fuel alcohol production from sugarcane.

2.5   Harvesting Plants for Bioenergy

Biomass can be converted into different types of products, including:

1. Electrical/heat energy
2. Transport fuel
3. Chemical feedstock

Woody and herbaceous species are the ones used most often by biomass researchers and industry. Several parameters are important in the biomass conversion process. The principal considerations in terms of the material type are moisture content, calorific value, fixed carbon and volatile proportion, ash/residue content, alkali metal content, and cellulose–lignin ratio. In a wet-biomass conversion process, the moisture content and cellulose–lignin ratio is of prime concern, while in a dry-biomass conversion process, it is the alkali metal content and cellulose–lignin ratio. The Laticiferous plant species of Apocyanaceae, Asclepiadaceae, Convolvulaceae, and Euphorbiaceae have been analyzed for use as renewable energy sources. Analysis of oil and hydrocarbon contents of 15 different plant species tested has revealed that *Carissa carandas* L., *Ceropegia juncea* Roxb., *Hemidesmus indicus* R. Br., and *Sarcostemma brunourianum* W. A. are the most suitable species [16]. In another study, five different plant species *Plumeria alba, C. procera, Euphorbia nerifolia, Nerium indicum,* and *Mimusops elengi* have been evaluated as potential renewable energy sources. Whole plants and plant parts (leaf, stem, and bark) have been analyzed for oil, polyphenol, hydrocarbons, crude protein, α-cellulose, lignin, ash, and mineral content. The barks of these plants were identified to have greater hydrocarbon content than the leaves. Based on the dry-biomass yields, hydrocarbon content, and other properties, these plant species most suitable for renewable energy sources have been identified [17]. In a study conducted on 51 plant species in Tennessee, in the United States, an examination of the oil, polyphenol, hydrocarbon, protein, and ash content reveals that *Lapsana communis* yields the maximum oil (6.1% dry, ash-free plant sample basis). *Chrysopsis graminifolia, Solidago erecta,* and *Verbesina alternifolia* have been identified as rubber-producing species with 0.4–0.7% hydrocarbon [18].

2.6 Products

Several processes similar to petroleum refining are involved in the conversion of biomass into different products. Biorefineries convert biomass into different products in different stages. The different stages involved in the conversion of biomass to products are depicted in Fig. 2.11.

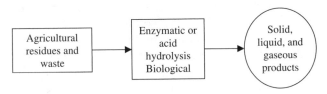

Figure 2.11 Different products from biomass.

There has been a tremendous increase in biobased products such as ethanol, high-fructose syrups, citric acid, monosodium glutamate, lysine, enzymes, and specialty chemicals worldwide. It is estimated that in 2000–2006 in United States alone, there will be an increase in the use of liquid fuels, organic chemicals, and biopolymers from the current level of ~2%, 10%, and 90% each to 10%, 25%, and 95%, respectively [19].

2.6.1   Gaseous products

In Chap. 1, gasification (pyrolysis) of biomass, biogas, gobargas, hydrogen, and biohydrogen were discussed in detail.

2.6.2   Liquid products

An important renewable energy resource for transportation purposes is liquid fuel based on plant oils. However, pure plant oils are generally not suitable for use in modern diesel engines. This can be overcome by the process of transesterification. The resultant fatty-acid methyl esters have properties similar to those of diesel and are commonly called biodiesel. Biodiesel presents several advantages, such as better CO_2 balance than diesel, low soot content, reduced hydrocarbon emissions, and low carcinogenic potential [20]. The specification standards for the European Union (EU) and the United States are EN14214 and ASTM D6751, respectively. The EU directive established a minimum content of 2% and 5.75% biodiesel for all petrol and diesel used in transport by December 31, 2005, and December 31, 2010, respectively. Biodiesel refers to the pure oil before blending with diesel fuel. Biodiesel blends are represented as "BXX," with "XX" representing the percentage of biodiesel component in the blend (National Biodiesel Board, 2005) [21]. In the biomass-to-liquid conversion processes, biomass is broken down into a gaseous constituent and a solid constituent by low-temperature gasification. The next step involves production of synthetic gas, which is converted into fuel (termed SunFuel) by the Fischer-Tropsch synthesis process, with downstream fuel optimization by hydrogen after treatment [22]. Ethanol has already been introduced in countries such as Brazil, the United States, and some European countries. In Brazil, it is currently produced from sugar and, in the United States, from starch at competitive prices. Ethanol is currently produced from sugarcane and starch-containing materials, where the conversion of starch to ethanol includes a liquefaction step (to make the starch soluble) and a hydrolysis step (to produce glucose). There are generally two types of processes for production of bioethanol: the lignocellulosic process and the starch process. Unlike the starch-based process, the lignocellulosic process has not been as widely adopted due to techno-economic reasons.

High ethanol yield requires complete hydrolysis of both cellulosic and hemicellulose with a minimum of sugar dehydration, followed by efficient fermentation of all sugars in the biomass. Certain advantages of using lignocellulose-based liquid biofuels are that they are evenly distributed across the globe and hence are readily available, less expensive compared to agricultural feedstock, produced at a lower cost, and have low net greenhouse gas emissions. Enzymatic processes (essentially using bacteria, yeasts, or filamentous fungi) have been considered for lignocellulosic processes. The enzymatic process when coupled with the fermentation process is known as simultaneous saccharification and fermentation. This has proved to be efficient in the fermentation of hexose and pentose sugars [23]. Genencor International (www.genencor.com/) and Novozymes, Inc., (www.novozymes.com) have been awarded $17 million each by the U.S. Department of Energy with a goal to reduce the enzyme cost tenfold (www.eere.energy.gov/). The Iogen Corp. (www.iogen.ca/) demo-plant is the only one that produces bioethanol from lignocellulose, using the enzymatic hydrolysis process. This plant is known to handle about 40 ton/day of wheat, oat, barley, and straw and is designed to produce up to 3 ML/yr of cellulose ethanol. Refer to Chap. 3 for bioethanol preparation, Chap. 6 for boidiesel processing, and Chap. 7 for ethanol and methanol used in engines.

2.6.3 Solid products

Refer to Sec. 1.14, Chap. 1, for more details on biomass. Solid products fall under the following categories:

1. Direct outcome of photosynthesis: Products from forest, shrubs, agricultures, and aquacultures.

2. Nonphotosynthesis: Mushrooms, animal biomass, indirect from photofixation.

3. Wastes: Forests and agricultural products.

4. Municipal solid wastes: Not all solid biomass may be suitable for different end uses, i.e., energy production or energy recovery. For example, mushrooms are notably useful as food, feed, or fodder, not otherwise. Biomass properties are guidelines to further and more fruitful end uses. The properties depend on the following:
   a. Water or moisture content (aqueous/dry)
   b. Calorific or combustion value
   c. Dry residues/ash content/silicates, and so forth
   d. Alkali metal/oxides in the ash
   e. Ratio of cellulose/liquid/oils/fats/of other carbonaceous matters
   f. Ratio of solid/liquid/volatiles

Direct combustion of biomass for heat generation is the most inefficient technique in energy economy, heat being the most inefficient of all forms of energy. The best way to utilize biomass is to recycle biomass for production of other or further biomass, namely, agriculture, horticulture, aquaculture, poultry, animal farming, and so forth. Randomness is reduced (low entropy change), and environmental chaos is lessened. Properties (a), (c), and (d) are significant for farming; (b) and (f) are important for hydrolytic processes; and (e) is important for biofuels and biodiesel. All the points are important for fermentations and in biorefineries. Biorefinery has become a new science and technology harmony for a promising future, which takes care of different aspects of biosafety, minimizes waste, and maximizes energy efficiency. It is a field of engineering and technology for the future. Biorefinery is a system similar to that of petroleum in its requirements for producing fuels and chemicals from biomass. A biorefinery is a capital-intensive project and is based on a conversion technology process of biomass. Hence, several technologies—thermochemical, chemical, biochemical, and so forth—are combined to reduce the overall cost. Fernando et al. suggest an integrated biorefinery process from bio-oil produced from pyrolysis of biomas (see Fig. 2.12),

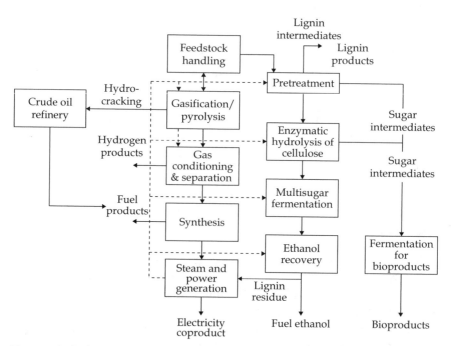

Figure 2.12   An integrated biorefinery process. (*Permission from S. Fernando, Associate Editor, FPEI—American Society of Agricultural and Biological Engineers (ASABE), Mississippi State University, USA.*)

which will not only produce sugar but also different by-products and electricity [24]. The process can produce its own power.

Fermentation is equally important. Anaerobic and restricted aerobic digestion with selected algae species allow us to harvest hydrogen and clean fuels, without much loss of biomass and with the least amount of waste products. In an aerobic process, the process is carried out by oxidizing the volatile matter into biodegradable organic fractions of solid waste. Air acts as a source of oxygen, and aerobic bacteria act as a catalyst. The change occurring during the process may be represented as

$$\text{Biomass} + O_2 \text{ (Aerobic bacteria)} \rightarrow CO_2 + H_2O + \text{Organic manure}$$

Anaerobic digestion is carried out by segregating the nonbiodegradables and the biodegradables at the same time. This may be done manually or mechanically. The smaller pieces of inorganic materials like clay and sand may be removed by washing the biomass with water. The washed material is then shredded into a size that will not interfere with mixing and may be more amenable to bacterial action. The shredded biomass is then mixed with sufficient quantity of water, and slurry is fed into a digester system. If necessary, nutrients like nitrogen, phosphorus, and potassium have to be added to the digester. The process involves four groups of bacteria in the digested slurry as follows:

1. Hydrolytic bacteria catabolize carbohydrates, proteins, lipids, and so forth contained in the biomass to fatty acids, H_2, and CO_2.
2. Hydrogen-producing acetogenic bacteria catabolize certain fatty acids and some neutral end products to acetate, CO_2, and H_2.
3. Homoacetogenic bacteria synthesize acetate, using H_2, CO_2, and formate.
4. In the final phase, called the methanogenic phase, methanogenic bacteria cleave acetate to methane and CO_2.

Water acts as a catalytic agent in methane formation. Thus water is acted upon by enzymes, itself breaking down to hydrogen and oxygen. Hydrogen is used by microorganisms to reduce CO_2 to CH_4, while oxygen oxidizes carbon dioxide, i.e., makes it acidic (H_2CO_3). In simple terms, acetate (in presence of CoI) is simultaneously oxidized to CO_2 and reduced to CH_4. For details, refer to Chap. 1, methanation, and Baker's and Ganzalus pathway. Thus, methane-forming bacteria play an important role in the circulation of substances and energy turnover in nature. They absorb CO, CO_2, and H_2 to give hydrocarbon and methane and help synthesis of their own cell substances. During anaerobic digestion, gas containing mainly CH_4 and CO_2 is produced. The gas is known as biogas, which is used for the generation of electricity or fuel. The residual biomass comes out of the digester in the form of a slurry, which is separated into a sludge, which is used as fertilizer and a stream of waste water. Research is ongoing to produce renewable energies from different plant sources, which will necessarily dominate the world's energy supply in the long-term. Using renewable-energy system technologies will create employment at much higher rates than any other technologies would [1]. There are economic opportunities for industries and craft jobs through production, installation, and maintenance of renewable energy systems.

References

1. European Renewable Energy Council (EREC). *Integration of Renewable Energy Sources: Targets and Benefits of Large-Scale Deployment of Renewable Energy Sources,* Workshop—Renewable Energy Market Development, Riga, Latvia, May 2004.
2. J. A. Bassham. Increasing crop production through more controlled photosynthesis, *Science* **197**, 630–638, 1977.
3. P. McKendry. Energy production from biomass (Part I): Overview of biomass, *Bioresource Technology* **83**, 37, 2002.
4. A. Nag. *Analytical Techniques in Agriculture, Biotechnology and Environmental Engineering,* New Delhi: Prentice-Hall, 2006.
5. R. E Häusler, H.-J. Hirsch, F. Kreuzaler, and C. Peterhänsel. Overexpression of C_4-cycle enzymes, *Journal of Experimental Botany* **53**(369), 591–607, 2002.
6. W. Amatayakul and C. Azaul. Eucalyptus for fossil fuel substitution and carbon sequestration: The costs of carbon dioxide abatement in Thailand, *International Journal of Sustainable Development* **6**(3), 359–377, 2003.
7. C. L. Schulze, E. Schnepf, and K. Motbes. Uber die Localisation der Kautschukpartikel in verschiedenen Typen von Milchróhren. Flora, *Abstracts* **158**, 458–460, 1967.
8. Energy Information Administration. Forecast and analysis of energy data, *International Energy Outlook 2005,* Report #: DOE/EIA-0484 (2005): www.eia.doe.gov/oiaf/ieo/oil.html.
9. E. Chlorent and R. P. Overend. Liquid fuels from lignocellulosics, In: *Biomass Regenerable Energy,* Hall, D. O. and Overend, R. P. (Eds.), Rochester, UK: John Wiley & Sons, pp. 257–269, 1987.
10. S. B. McLaughlin, R. Samson, D. Bransby, and A. Wiselogel. Evaluating physical, chemical, and energetic properties of perennial grasses as biofuels, In: *Proceedings of the Seventh National Bioenergy Conference—Bioenergy '96,* Nashville, TN, September 15–20, 1996.
11. L. Dewey. *Hemp Hurds as Papermaking Material,* USDA Bulletin No. 404, US Government Printing Office, Washington, DC, October 14, 1916.
12. R. Palaniraj and S. C. Sati. Evaluation of of *Artocarpus hirsute* and *Ficus elastica* as renewable source of energy, *Indian Journal of Agricultural Chemistry* **36**(1), 23, 2003.
13. B. K. Behera, M. Arora, and D. K. Sharma. Studies on biotransformation of *Calotropis procera* latex–A renewable source of petroleum, value-added chemicals, and products, *Energy Sources* **22**(9), 781, 2000.
14. Ove Arup and Partners. *Monitoring of a Commercial Demonstration of Harvesting and Combustion of Forestry Wastes,* ETSU B/1171-P1, London, UK, 1989.
15. F. Culshaw and C. Butler. *A Review of the Potential of Bio-Diesel as a Transport Fuel,* ETSU-R-71, The Stationary Office, London, UK, 1992.
16. T. Sekar and K. Francis. Some plant species screened for energy, hydrocarbons and phytochemicals, *Bioresource Technology* **65**(3), 257–259, 1998.
17. D. Kalita and C. N. Saikia. Chemical constituents and energy content of some latex bearing plants, *Bioresource Technology* **92**(3), 219–227, 2004.
18. M. E. Carr and M. O. Bagby. Tennessee plant species screened for renewable energy sources, *Economic Botany* **41**(1), 78–85, 1987.
19. S. Fernando, C. Hall, and S. Jha. NO_x reduction from biodiesel fuels, *Energy Fuels* **20**, 376–382, 2006.

20. G. Vicente, M. Martinez, and J. Aracil. Kinetics of *Brassica carinata* oil methanolysis, *Energy Fuels* **20**, 1722–1726, 2006.

21. W. Steiger. Biomass-to-liquid fuels: Energy for future, In: *Proceedings of World Renewable Energy Congress VIII: Linking the World with Renewable Energy*, Denver, co, August 29–September, 2004, 32–33.

22. B. Hahn-Hagerdal, S. Galbe, G. Gorwa-Grauslund, and Z. G. Liden. The fuel of tomorrow from the residues of today, *Trends in Biotechnology* **24**(12), 549–556, 2006.

23. National Biodiesel Board. Biodiesel 101: www.biodiesel.org/resources/biodiesel_basics/default.shtm.

24. S. Fernando, S. Adhikari, C. Chandrapal, and N. Murali. Biorefineries: Current status, challenges and future direction, *Energy and Fuels* **20**, 1727–1737, 2006.

Bioethanol: Market and Production Processes

Mohammad J. Taherzadeh and Keikhosro Karimi

3.1 Introduction

Ethanol (C_2H_5OH) is a clear, colorless, flammable chemical. It has been produced and used as an alcoholic beverage for several thousand years. Ethanol also has several industrial applications (e.g., in detergents, toiletries, coatings, and pharmaceuticals) and has been used as transportation fuel for more than a century. Nicholas Otto used ethanol in the internal combustion engine invented in 1897 [1]. However, ethanol did not have a major impact in the fuel market until the 1970s, when two oil crises occurred in 1973 and 1979. Since the 1980s, ethanol has been a major actor in the fuel market as an alternative fuel as well as an oxygenated compound for gasoline. Ethanol can be produced synthetically from oil and natural gas, or biologically from sugar, starch, and lignocellulosic materials. The biologically produced ethanol is sometimes called *fermentative ethanol* or *bioethanol*. Application of bioethanol as fuel has no or very limited net emission of CO_2 [2] and is able to fulfill the Kyoto Climate Change Protocol (1997) to decrease the net emission of CO_2 [3]. In this chapter, the global market and the production of bioethanol are briefly reviewed.

3.2 Global Market of Bioethanol and Future Prospects

Ethanol is produced from a variety of feedstocks. Fermentative ethanol is produced from grains, molasses, sugarcane juice, fruits, surplus wine, whey, and some other similar sources, which contain simple sugars and their polymers. On the other hand, synthetic ethanol is produced from oil, e.g., through hydration of ethylene:

$$\text{Oil} \rightarrow CH_2 = CH_2 \text{ (ethylene)} \xrightarrow{H_2O} CH_3CH_2OH \text{ (ethanol)} \qquad (3.1)$$

Several companies, such as Sasol, SADAF, British Petroleum, and Equistar, produce synthetic ethanol, with capacities of 100–400 kilotons/yr. However, the share of synthetic ethanol in world ethanol production was less than 4% in 2006, down from 7% in the 1990s [4]. Furthermore, increasing oil price or declining ethanol price can harm the economic competition of synthetic ethanol production, compared to the fermentative one. Ethylene prices in 2005 rose to US $1000 per ton, while ethanol values were around US $500 per ton. If we consider the theoretical yield of ethanol from ethylene based on Eq. (3.1) as 1.64 kg/kg, the price of raw materials was higher than that of the product. In this case, it is economically feasible to produce biobased plastics through "bioethylene":

$$\text{Biomass/crops} \xrightarrow{\text{Fermentation}} CH_3CH_2OH \xrightarrow{H_2O} CH_2 = CH_2 \rightarrow \text{Plastics} \quad (3.2)$$

The global demand for ethylene is around 120 megatons [4]. It can be considered a new market for ethanol in the future.

The total world ethanol production in 2006 was 49.8 GL (gigaliter) (39 megatons), where 77% of this production was used as fuel, 8% as beverage, and 15% in industrial applications [4]. Since 1975, potable ethanol production has not experienced a major growth, while industrial ethanol production has experienced growth by about 75%. However, fuel ethanol production has increased aggressively from less than 1 GL in 1975 to more than 38 GL in 2006 (see Fig. 3.1).

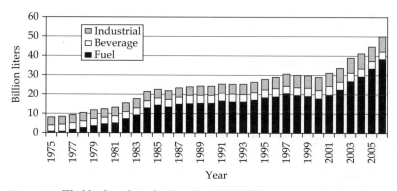

Figure 3.1   World ethanol production since 1976 [4].

There is competition between Brazil and the United States to be the dominant ethanol producer in the world. So far, Brazil has been the largest ethanol producer, but the statistics from 2006 imply that the United States is the largest ethanol producer with 19.1 GL, followed by Brazil with 16.7 GL. Both countries produced almost identical amounts of ethanol in 2005 (16.2 and 16.0 GL, respectively). The American continents produced 72% of the world ethanol production (see Fig. 3.2), followed by Asia, Europe, Oceania, and the African continents.

There is tough competition between sugar crops (particularly sugarcane juice and molasses) and starch crops (particularly maize) as feedstock for fuel ethanol production. While sugar crops were the feedstock for more than 60% of fuel ethanol production at the beginning of the 2000s, its share decreased to 47% in 2006 and starch crops were used for 53% of fuel ethanol production in the same year.

The world fuel ethanol production is predicted to keep the latest trend, at least until 2015. In comparison to 2006, ethanol production by Brazil and the United States is expected to increase by 102% and 93%, respectively. However, total production of the rest of the world is expected to increase by 585% [4]. Therefore, the world fuel ethanol production is expected to increase to around 100 GL. The main reasons for this sharp increase in ethanol production and demand in the future might be [2, 5, 6]:

■ Possible increase in oil prices

■ Higher demand for liquid fuels in the future

■ Decline of the crude oil supply in the future

■ Environmental legislation in different countries to encourage using biofuels

■ Production of bioplastic materials from ethanol

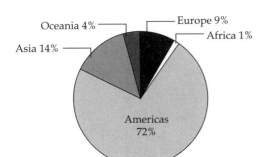

Figure 3.2 World ethanol production in 2006 divided by continents [4].

3.3 Overall Process of Bioethanol Production

The process of ethanol production depends on the raw materials used. A general simplified representation of these processes is shown in Fig. 3.3, and a brief description of different units of the process is presented in the rest of this chapter. It should be noted that if sugar substances, such as molasses and sugarcane juice, are used as raw materials, then milling, pretreatment, hydrolysis, and detoxification are not necessary.

Milling, liquefaction, and saccharification processes are usually necessary for production of fermentable sugar from starchy materials, while milling, pretreatment, and hydrolysis are typically used for ethanol production from lignocellulosic materials. Furthermore, a detoxification unit is not always considered, unless a toxic substrate is fed to the bioreactors.

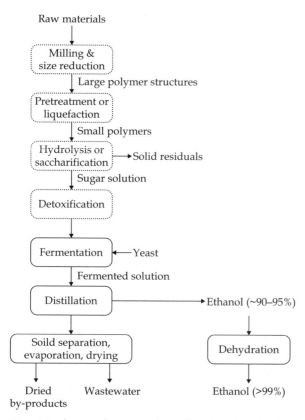

Figure 3.3 A general process scheme for ethanol production from different raw materials.

3.4   Production of Sugars from Raw Materials

Sugar substances (such as sugarcane juice and molasses), starchy materials (such as wheat, corn, barley, potato, and cassava), and lignocellulosic materials (such as forest residuals, straws, and other agricultural by-products) are being considered as the raw materials for ethanol production. The dominating sugars available or produced from these popular raw materials are

- Glucose, fructose, and sucrose in sugar substances
- Glucose in starchy materials
- Glucose from cellulose and either mannose or xylose from hemicellulose of lignocellulosic materials

Most ethanol-producing microorganisms can utilize a variety of hexoses such as glucose, fructose, galactose, and mannose, and a limited number of disaccharides such as sucrose, lactose, cellobiose, and maltose, and rarely their polymers. Therefore, it is necessary to convert the complex polysaccharides, such as cellulose and starch, to simple sugars or disaccharides. Different types of substrates that need treatment are presented in Table 3.1, prior to fermentation.

In this section, sugar production from starchy materials is discussed; lignocellulosic materials are discussed in Sec. 3.5.

3.4.1   Sugar solution from starchy materials

There are various raw materials that contain starch and are suitable for ethanol production. Corn is the most widely used on an industrial scale for this purpose. However, there are several other cereals, such as wheat, rye, barley, and sorghum, and crop roots such as potato and cassava, which are used as raw materials for ethanol production. The cereals contain about 60–70% starch, 8–12% proteins, 10–15% water, and small

TABLE 3.1   Treatment for Different Types of Substrates

Substrate	Pretreatment or liquefaction	Hydrolysis or saccharification	Detoxification
Potential sugar substrates	No	No	Typically no
Starchy materials	Yes	Yes	No
Lignocellulose materials	Yes	Yes	Depends on the hydrolysis method

amounts of fats and fibers. The compositions of the crop roots are almost identical to those of the cereals on a dry basis, but the water content of the roots is usually 70–80%. The exact composition of each raw material depends on the type and variety of materials used and can be found in literature (e.g., [7]). Starch from these materials is used as a carbon and energy source, and part of the proteins as a nitrogen source, by the microorganisms.

Starch contains two fractions: amylose and amylopectin. Amylose, which typically constitutes about 20% of starch, is a straight-chain polymer of α-glucose subunits with a molecular weight that may vary from several thousands to half a million. Amylose is a water-insoluble polymer. The bulk of starch is amylopectin, which is also a polymer of glucose. Amylopectin contains a substantial number of branches in the molecular chains. Branches occur from the ends of amylose segments, averaging 25 glucose units in length. Amylopectin molecules are typically larger than amylose, with molecular weights ranging up to 1–2 Mg. Amylopectin is soluble in water and can form a gel by absorbing water.

For ethanol production, hydrolysis is necessary for converting starch into fermentable sugar available to microorganisms. Traditional conversion of starch into sugar monomers requires a two-stage hydrolysis process: liquefaction of large starch molecules to oligomers, and saccharification of the oligomers to sugar monomers. This hydrolysis may be catalyzed by acid or amylolytic enzymes.

3.4.2   Acid hydrolysis of starch

Acid hydrolysis is an old process still applied in some ethanol industries. Sulfuric acid is the most commonly applied acid in this process, where starch is converted to low-molecular-weight dextrins and glucose [8]. Main advantages of this process are rapid hydrolysis and less cost for catalyst, compared to the enzymatic hydrolysis. However, the acid processes possess drawbacks including (a) high capital cost for an acid-resistant hydrolysis reactor, (b) destruction of sensitive nutrients such as vitamins present in raw materials, and (c) further degradation of sugar to hydroxymethylfurfural (HMF), levulinic acid, and formic acid, which lowers the ethanol yield and inhibits the fermentation process [9].

The acid hydrolysis process can be performed either in batch or in continuous systems. Dilute-acid hydrolysis can also be used as a pretreatment for enzymatic hydrolysis. It is common to soak the starch or starchy materials in the dilute acid prior to enzymatic hydrolysis, then to continuously pass it through a steam-jet heater into a cooking tube (called a jet cooker or mash cooker) with a plug flow residence time for a couple of minutes, and then subject it to enzymatic hydrolysis.

3.4.3 Enzymatic hydrolysis of starch

Enzymatic hydrolysis has several advantages compared to acid hydrolysis. First, the specificity of enzymes allows the production of sugar syrups with well-defined physical and chemical properties. Second, milder enzymatic hydrolysis results in few side reactions and less "browning" [8]. Different types of enzymes involved in the enzymatic hydrolysis of starch are α-amylase, β-amylase, glucoamylase, pulluanases, and isoamylases. The mechanism of action of these enzymes is presented schematically in Fig. 3.4.

There are two popular industrial processes from starch materials, dry milling and wet milling. In the dry-milling process, grain is first ground into flour and then processed without separation of the starch from germ and fiber components. In this method, the mixture of starch and other components is processed. Starch is converted to sugar in two stages: liquefaction and saccharification, by adding water, enzymes, and heat (enzymatic hydrolysis). Dry-milling processes produce a coproduct, distillers' dried grains with solubles (DDGS), which is used as an animal-feed supplement. Without the revenues from that coproduct, ethanol from dry-milled corn processing would not be economically favorable [2]. A dry-milling process for alcohol production processes the whole grain, or components derived from the whole grain. Saccharification and fermentation of dry-milled corn result in ethanol and distillers' dried grains (DDG). When DDG are combined with fermentation liquids and dried, they result in DDGS as the major feed by-product [10].

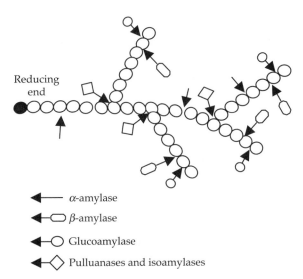

Reducing
end

◀━━━ α-amylase
◀━◯ β-amylase
◀━◯ Glucoamylase
◀━◇ Pulluanases and isoamylases

Figure 3.4 Mechanism of action of amylase on starch.

In the wet-milling process, grain is steeped and separated into starch, germ, and fiber components. Wet milling is capital intensive, but it generates numerous coproducts that help to improve the overall production economics [2]. Wet mills produce corn gluten feed, corn gluten meal, corn germ, and other related coproducts. In this method, after the grain is cleaned, it is steeped and then ground to remove the germ. Further grinding, washing, and filtering steps separate the fiber and gluten. The starch that remains after these separation steps is then broken down into fermentable sugars by the addition of enzymes in the liquefaction and saccharification stages. The fermentable sugars produced are then subjected to fermentation for ethanol production, like the other fermentable sugars.

3.5   Characterization of Lignocellulosic Materials

Lignocellulosic materials predominantly contain a mixture of carbohydrate polymers (cellulose and hemicellulose) and lignin. The carbohydrate polymers are tightly bound to lignin mainly through hydrogen bonding, but also through some covalent bonding. The contents of cellulose, hemicellulose, and lignin in common lignocellulosic materials are listed in Table 3.2. Different types of carbohydrates (glucan, xylan, galactan, arabinan, and mannan), lignin, extractive, and ash content of many lignocellulosic materials have been analyzed and are available in the literature [2, 11–14] (see Table 3.2).

3.5.1   Cellulose

Cellulose is the main component of most lignocellulosic materials. Cellulose is a linear polymer of up to 27,000 glucosyl residues linked by β-1,4 bonds. However, each glucose residue is rotated 180° relative to

TABLE 3.2   Contents of Cellulose, Hemicellulose, and Lignin in Common Lignocellulosic Materials

Lignocellulosic materials	Cellulose (%)	Hemicellulose (%)	Lignin (%)
Hardwood stems	40–75	10–40	15–25
Softwood stems	30–50	25–40	25–35
Corn cobs	45	35	15
Wheat straw	30	50	15
Rice straw	32–47	19–27	5–24
Sugarcane bagasse	40	24	25
Leaves	15–20	80–85	0
Paper	85–99	0	0–15
Newspaper	40–55	25–40	18–30
Waste paper from chemical pulps	60–70	10–20	5–10
Grasses	25–40	25–50	10–30

its neighbors so that the basic repeating unit is in fact cellobiose, a dimer of a two-glucose unit. As glucose units are linked together into polymer chains, a molecule of water is lost, which makes the chemical formula $C_6H_{10}O_5$ for each monomer unit of "glucan." The parallel polyglucan chains form numerous intra- and intermolecular hydrogen bonds, which result in a highly ordered crystalline structure of native cellulose, interspersed with less-ordered amorphous regions [15, 16].

3.5.2 Hemicellulose

Hemicelluloses are heterogeneous polymers of pentoses (e.g., xylose and arabinose), hexoses (e.g., mannose, glucose, and galactose), and sugar acids. Unlike cellulose, hemicelluloses are not chemically homogeneous. Hemicelluloses are relatively easily hydrolyzed by acids to their monomer components consisting of glucose, mannose, galactose, xylose, arabinose, and small amounts of rhamnose, glucuronic acid, methylglucuronic acid, and galacturonic acid. Hardwood hemicelluloses contain mostly xylans, whereas softwood hemicelluloses contain mostly glucomannans. Xylans are the most abundant hemicelluloses. Xylans of many plant materials are heteropolysaccharides with homopolymeric backbone chains of 1,4-linked β-D-xylopyranose units. Xylans from different sources, such as grasses, cereals, softwood, and hardwood, differ in composition. Besides xylose, xylans may contain arabinose, glucuronic acid or its 4-*O*-methyl ether, and acetic, ferulic, and *p*-coumaric acids. The degree of polymerization of hardwood xylans (150–200) is higher than that of softwoods (70–130) [14, 15].

3.5.3 Lignin

Lignin is a very complex molecule. It is an aromatic polymer constructed of phenylpropane units linked in a three-dimensional structure. Generally, softwoods contain more lignin than hardwoods. Lignins are divided into two classes, namely, "guaiacyl lignins" and "guaiacyl-syringyl lignins." Although the principal structural elements in lignin have been largely clarified, many aspects of their chemistry remain unclear. Chemical bonds have been reported between lignin and hemicellulose, and even cellulose. Lignins are extremely resistant to chemical and enzymatic degradation. Biological degradation can be achieved mainly by fungi, but also by certain actinomycetes [15, 17].

3.6 Sugar Solution from Lignocellulosic Materials

There are several possible ways to hydrolyze lignocellulose (see Fig. 3.5). The most commonly applied methods can be classified into two groups: chemical hydrolysis and enzymatic hydrolysis. In addition, there are

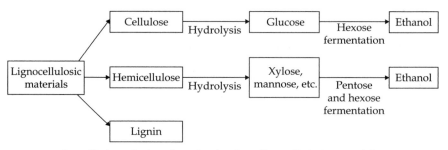

Figure 3.5 Overall view of ethanol production from lignocellulosic materials.

some other hydrolysis methods in which no chemicals or enzymes are applied. For instance, lignocellulose may be hydrolyzed by γ-ray or electron-beam irradiation, or microwave irradiation. However, those processes are commercially unimportant [15].

3.6.1 Chemical hydrolysis of lignocellulosic materials

Chemical hydrolysis involves exposure of lignocellulosic materials to a chemical for a period of time, at a specific temperature, and results in sugar monomers from cellulose and hemicellulose polymers. Acids are predominantly applied in chemical hydrolyses. Sulfuric acid is the most investigated acid, although other acids such as hydrochloric acid (HCl) have also been used. Acid hydrolyses can be divided into two groups: concentrated-acid hydrolysis and dilute-acid hydrolysis [18].

Concentrated-acid hydrolysis. Hydrolysis of lignocellulose by concentrated sulfuric or hydrochloric acids is a relatively old process. Concentrated-acid processes are generally reported to give higher sugar and ethanol yield, compared to dilute-acid processes. Furthermore, they do not need a very high pressure and temperature. Although this is a successful method for cellulose hydrolysis, concentrated acids are toxic, corrosive, and hazardous, and these acids require reactors that are highly resistant to corrosion. High investment and maintenance costs have greatly reduced the commercial potential for this process. In addition, the concentrated acid must be recovered after hydrolysis to make the process economically feasible. Furthermore, the environmental impact strongly limits the application of hydrochloric acid [12, 15].

Dilute-acid hydrolysis. Dilute-sulfuric acid hydrolysis is a favorable method for either the pretreatment before enzymatic hydrolysis or the conversion of lignocellulose to sugars. This pretreatment method gives high reaction rates and significantly improves enzymatic hydrolysis.

Depending on the substrate used and the conditions applied, up to 95% of the hemicellulosic sugars can be recovered by dilute-acid hydrolysis from the lignocellulosic feedstock [2, 13]. Of all dilute-acid processes, the processes using sulfuric acid have been the most extensively studied. Sulfuric acid is typically used in 0.5–1.0% concentration. However, the time and temperature of the process can be varied. It is common to use one of the following conditions in dilute-acid hydrolysis:

- Mild conditions, i.e., low pressure and long retention time
- Severe conditions, i.e., high pressure and short retention time

In dilute-acid hydrolysis, the hemicellulose fraction is depolymerized at temperatures lower than the cellulose fraction. If higher temperature or longer retention times are applied, the monosaccharides formed will be further hydrolyzed to other compounds. It is therefore suggested that the hydrolysis process be carried out in at least two stages. The first stage is carried out at relatively milder conditions during which the hemicellulose fraction is hydrolyzed, and a second stage can be carried out by enzymatic hydrolysis or dilute-acid hydrolysis, at higher temperatures, during which the cellulose is hydrolyzed [13]. These first and second stages are sometimes called "pretreatment" and "hydrolysis," respectively.

Hydrolyzates of first-stage dilute-acid hydrolysis usually consist of hemicellulosic carbohydrates. The dominant sugar in the first-stage hydrolyzate of hardwoods (such as alder, aspen, and birch) and most agricultural residues such as straw is xylose, whereas first-stage hydrolyzates of softwoods (e.g., pine and spruce) predominantly contain mannose. However, the dominant sugar in the second-stage hydrolyzate of all lignocellulosic materials, either by enzymatic or dilute-acid hydrolysis, is glucose, which originates from cellulose.

Detoxification of acid hydrolyzates. In addition to sugars, several by-products are formed or released in the acid hydrolysis process. The most important by-products are carboxylic acids, furans, and phenolic compounds (see Fig. 3.6).

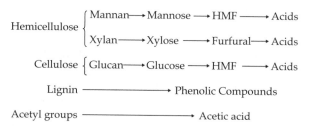

Figure 3.6 Formation of inhibitory compounds from lignocellulosic materials during acid hydrolysis.

Acetic acid, formic acid, and levulinic acid are the most common carboxylic acids found in hydrolyzates. Acetic acid is mainly formed from acetylated sugars in the hemicellulose, which are cleaved off already at mild hydrolysis conditions. Since the acid is not further hydrolyzed, formation of acetic acid is dependent on the temperature and pressure of dilute-acid hydrolysis, until the acetyl groups are fully hydrolyzed. Therefore, the acetic acid yield in the hydrolysis does not significantly depend on the severity of the hydrolysis process [13, 19].

Furfural and HMF are the only furans usually found in hydrolyzates in significant amounts. They are hydrolysis products of pentoses and hexoses, respectively [13]. Formation of these by-products is affected by the type and size of lignocellulose, as well as hydrolysis variables such as acid type and concentration, pressure and temperature, and the retention time.

A large number of phenolic compounds have been found in hydrolyzates. However, reported concentrations are normally a few milligrams per liter. This could be due to the low water solubility of many of the phenolic compounds, or a limited degradation of lignin during the hydrolysis process. Vanillin, syringaldehyde, hydroxybenzaldehyde, phenol, vanillic acid, and 4-hydroxybenzoic acid are among the phenolic compounds found in dilute-acid hydrolyzates [18].

Biological (e.g., using enzymes peroxidase and laccase), physical (e.g., evaporation of volatile fraction and extraction of nonvolatile fraction by diethyl ether), and chemical (e.g., alkali treatment) methods have been employed for detoxification of lignocellulosic hydrolyzates [20, 21].

Detoxification of lignocellulosic hydrolyzates by overliming is a common method used to improve fermentability [22–25]. In this method, $Ca(OH)_2$ is added to hydrolyzates to increase the pH (up to 9–12) and keep this condition for a period of time (from 15 min up to several days), followed by decreasing the pH to 5 or 5.5. Recently, it has been found that time, pH, and temperature of overliming are the effective parameters in detoxification [26]. However, the drawback of this treatment is that part of the sugar is also degraded during the overliming process. Therefore, it is necessary to optimize the process to achieve a fermentable hydrolyzate without any loss of the sugar [21, 26].

3.6.2 Pretreatment prior to enzymatic hydrolysis of lignocellulosic materials

Native (indigenous) cellulose fractions of cellulosic materials are recalcitrant to enzymatic breakdown, so a pretreatment step is required to render them amenable to enzymatic hydrolysis to glucose. A number of pretreatment processes have been developed in laboratories, including:

■ *Physical pretreatment*—mechanical comminution, irradiation, and pyrolysis

- *Physicochemical pretreatment*—steam explosion or autohydrolysis, ammonia fiber explosion (AFEX), SO_2 explosion, and CO_2 explosion
- *Chemical pretreatment*—ozonolysis, dilute-acid hydrolysis, alkaline hydrolysis, organosolvent process, and oxidative delignification
- *Biological pretreatment*

However, not all of these methods may be technically or economically feasible for large-scale processes. In some cases, a method is used to increase the efficiency of another method. For instance, milling could be applied to achieve better steam explosion by reducing the chip size. Furthermore, it should be noticed that the selection of pretreatment method should be compatible with the selection of hydrolysis. For example, if acid hydrolysis is to be applied, a pretreatment with alkali may not be beneficial [18]. Pretreatment methods have been reviewed by Wyman [2] and Sun and Cheng [12].

Among the different types of pretreatment methods, dilute-acid, SO_2, and steam explosion methods have been successfully developed for pretreatment of lignocellulosic materials. The methods show promising results for industrial application. Dilute-sulfuric acid hydrolysis is a favorable method for either pretreatment before enzymatic hydrolysis or conversion of lignocellulose to sugars.

3.6.3 Enzymatic hydrolysis of lignocellulosic materials

Enzymatic hydrolysis of cellulose and hemicellulose can be carried out by highly specific cellulase and hemicellulase enzymes (glycosyl hydrolases). This group includes at least 15 protein families and some subfamilies [15, 27]. Enzymatic degradation of cellulose to glucose is generally accomplished by synergistic action of three distinct classes of enzymes [2]:

- 1,4-β-D-glucan-4-glucanohydrolases or Endo-1,4-β-glucanases, which are commonly measured by detecting the reducing groups released from carboxymethylcellulose (CMC).
- Exo-1,4-β-D-glucanases, including both 1,4-β-D-glucan hydrolases and 1,4-β-D-glucan cellobiohydrolases. 1,4-β-D-glucan hydrolases liberate D-glucose and 1,4-β-D-glucan cellobiohydrolases liberate D-cellobiose.
- β-D-glucoside glucohydrolases or β-D-glucosidases, which release D-glucose from cellobiose and soluble cellodextrins, as well as an array of glycosides.

There is a synergy between exo–exo, exo–endo, and endo–endo enzymes, which has been demonstrated several times.

Substrate properties, cellulase activity, and hydrolysis conditions (e.g., temperature and pH) are the factors that affect the enzymatic hydrolysis of cellulose. To improve the yield and rate of enzymatic hydrolysis, there has been some research focused on optimizing the hydrolysis process and enhancing cellulase activity. Substrate concentration is one of the main factors that affect the yield and initial rate of enzymatic hydrolysis of cellulose. At low substrate levels, an increase of substrate concentration normally results in an increase of the yield and reaction rate of the hydrolysis. However, high substrate concentration can cause substrate inhibition, which substantially lowers the rate of hydrolysis, and the extent of substrate inhibition depends on the ratio of total substrate to total enzyme [12].

Increasing the dosage of cellulases in the process to a certain extent can enhance the yield and rate of hydrolysis, but would significantly increase the cost of the process. Cellulase loading of 10 FPU/g (filter paper units per gram) of cellulose is often used in laboratory studies because it provides a hydrolysis profile with high levels of glucose yield in a reasonable time (48–72 h) at a reasonable enzyme cost. Cellulase enzyme loadings in hydrolysis vary from 5 to 33 FPU/g substrate, depending on the type and concentration of substrates. β-glucosidase acts as a limiting agent in enzymatic hydrolysis of cellulose. Adding supplemental β-glucosidase can enhance the saccharification yield [28, 29].

Enzymatic hydrolysis of cellulose consists of three steps [12]: (1) adsorption of cellulase enzymes onto the surface of cellulose, (2) biodegradation of cellulose to simple sugars, and (c) desorption of cellulase. Cellulase activity decreases during hydrolysis. Irreversible adsorption of cellulase on cellulose is partially responsible for this deactivation. Addition of surfactants during hydrolysis is capable of modifying the cellulose surface property and minimizing the irreversible binding of cellulase on cellulose. Tween-20 and Tween-80 are the most efficient nonionic surfactants in this regard. Addition of Tween-20 as an additive in simultaneous saccharification and fermentation (SSF) at 2.5 g/L has several positive effects in the process. It increases the ethanol yield, increases the enzyme activity in the liquid fraction at the end of the process, reduces the amount of enzyme loading, and reduces the required time to attain maximum ethanol concentration [30].

3.7   Basic Concepts of Fermentation

The general reaction for ethanol production during fermentation is

$$\text{Sugar(s)} \xrightarrow{\text{Microorganisms}} \text{Ethanol + By-products}$$

In this reaction, the microorganisms work as a catalyst.

3.8   Conversion of Simple Sugars to Ethanol

Conversion of simple hexose sugars, such as glucose and mannose, in fermentation into ethanol can take place anaerobically as follows:

$$C_6H_{12}O_6 \text{ (Hexoses)} \xrightarrow{\text{Microorganisms}} 2C_2H_5OH \text{ (ethanol)} + 2CO_2$$

If the entire sugar is converted into ethanol according to the above reaction, the yield of ethanol will be 0.51 g/g of the consumed sugars, meaning that from 1.0 g of glucose, 0.51 g of ethanol can be produced. This is the theoretical yield of ethanol from hexoses. However, the ethanol yield obtained in fermentation does not usually exceed 90–95% of the theoretical yield, since part of the carbon source in sugars is converted to biomass of the microorganisms and other by-products such as glycerol and acetic acid [9, 31].

A similar reaction for anaerobic conversion of pentoses, such as xylose to ethanol, might be considered. Xylose is generally converted first to xylulose by a one-step reaction catalyzed by xylose isomerase (XI) in many bacteria, or by a two-step reaction through xylitol in yeasts and fungi. It can then be converted to ethanol anaerobically through a pentose phosphate pathway (PPP) and glycolysis. The general reaction can be written as

$$3C_5H_{10}O_5 \text{ (Pentoses)} \xrightarrow{\text{Microorganisms}} 5C_2H_5OH \text{ (Ethanol)} + 5CO_2$$

In this case, we can expect a theoretical ethanol yield of 0.51 g/g from xylose, as we had from glucose. However, the redox imbalance and slow rate of ATP formation are two major factors that make anaerobic ethanol production from xylose very difficult [32, 33]. A few anaerobic ethanol-producing strains have been developed from xylose in research laboratories, but no strain is so far available for industrial-scale processes. Attempts have been made to overcome this problem of xylose assimilation by cometabolization or working with microaerobic conditions, where oxygen is available at low concentrations. A number of microorganisms can produce ethanol aerobically from xylose, where the practical yield of ethanol from xylose and other pentoses is usually lower than its theoretical yield. The challenges in ethanol production from xylose have been reviewed by van Maris et al. [34].

3.9   Biochemical Basis of Ethanol Production from Hexoses

A simplified central metabolic pathway for ethanol production in yeast and bacteria under anaerobic conditions is presented in Fig. 3.7 [15, 35–37].

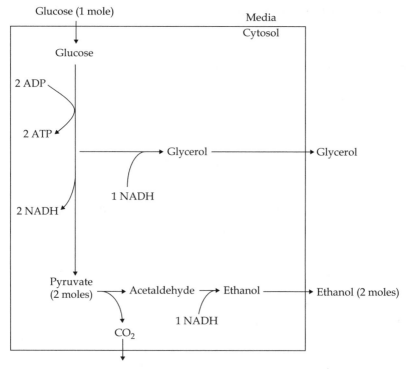

Figure 3.7 Central metabolic pathway in yeast under anaerobic conditions.

Three major interrelated pathways that control catabolism of carbohydrate in most ethanol-producing organisms are

■ Embden-Meyerhof pathway (EMP) or glycolysis
■ Pentose phosphate pathway (PPP)
■ Krebs or tricarboxylic acid cycle (TCA)

In glycolysis, glucose is anaerobically converted to pyruvic acid and then to ethanol through acetaldehyde. This pathway provides energy in the form of ATP to the cells. The net yield in glycolysis is 2 moles of pyruvate (or ethanol) and 2 moles of ATP from each mole of glucose. This pathway is also the entrance of other hexoses such as fructose, mannose, and galactose to metabolic pathways. With only 2 moles of ATP formed per glucose catabolized, large amounts of ethanol (at least 3.7 g of ethanol per gram of biomass) must be formed [15, 38].

The PPP handles pentoses and is important for nucleotide (ribose-5-phosphate) and fatty acid biosynthesis. The PPP is mainly used to reduce $NADP^+$. In *Saccharomyces cerevisiae,* 6–8% of glucose passes through the PPP under anaerobic conditions [8, 15].

The TCA cycle functions to convert pyruvic and lactic acids and ethanol aerobically to the end products CO_2 and H_2O. It is also a common channel for the ultimate oxidation of fatty acids and the carbon skeletons of many amino acids. In cells containing the additional aerobic pathways, the NADH that forms during glycolysis results in ATP generation in the TCA cycle [8].

Ethanol production from hexoses is redox-neutral, i.e., no net formation of NADH or NADPH occurs. However, biosynthesis of the cells results in net formation of NADH and consumption of NADPH. The PPP is mainly used to reduce $NADP^+$ to NADPH. Oxidation of surplus NADH under anaerobic conditions in *S. cerevisiae* is carried out through the glycerol pathway. Furthermore, there are other by-products—mainly carboxylic acids: acetic acid, pyruvic acid, and succinic acid—that add to the surplus NADH. Consequently, glycerol is also formed to compensate the NADH formation coupled with these carboxylic acids. Thus, formation of glycerol is coupled with biomass and carboxylic acid formation in anaerobic growth of *S. cerevisiae* [15, 39].

We should keep in mind that growth of the cells and increasing their biomass is the ultimate goal of the cells. They produce ethanol under anaerobic conditions in order to provide energy through catabolic reactions. Glycerol is formed to keep the redox balance of the cells, and carboxylic acids may leak from the cells to the medium. Therefore, the ethanol-producing microorganisms produce ethanol as the major product under anaerobic conditions, while biomass, glycerol, and some carboxylic acids are the by-products.

3.10   Chemical Basis of Ethanol Production from Pentoses

In general, yeast and filamentous fungi metabolize xylose through a two-step reaction before they enter the central metabolism (glycolysis) through the PPP. The first step is conversion of xylose to xylitol using xylose reductase (XR), and the second step is conversion of xylitol to xylulose using another enzyme, xylitol dehydrogenase (XDH) [40–42].

Wild strains of *S. cerevisiae* possess the enzymes XR and XDH, but their activities are too low to allow growth on xylose. Although *S. cerevisiae* cannot utilize xylose, it can utilize its isomer, xylulose. Thus, if *S. cerevisiae* is to be used for xylose fermentation, it requires a genetic modification to encode XR/XDH or XI [40, 43].

Bacteria have a slightly different metabolic pathway for xylose utilization. They convert xylose to xylulose in one reaction using XI [10, 44–46].

3.11 Microorganisms Related to Ethanol Fermentation

The criteria for an ideal ethanol-producing microorganism are to have (a) high growth and fermentation rate, (b) high ethanol yield, (c) high ethanol and glucose tolerance, (d) osmotolerance, (e) low optimum fermentation pH, (f) high optimum temperature, (g) general hardiness under physiological stress, and (h) tolerance to potential inhibitors present in the substrate [31, 47]. Ethanol and sugar tolerance allows the conversion of concentrated feeds to concentrated products, reducing energy requirements for distillation and stillage handling. Osmotolerance allows handling of relatively dirty raw materials with their high salt content. Low-pH fermentation combats contamination by competing organisms. High temperature tolerance simplifies fermentation cooling. General hardiness allows microorganisms to survive stress such as that of handling (e.g., centrifugation) [47]. The microorganisms should also tolerate the inhibitors present in the medium.

3.11.1 Yeasts

Historically, yeasts have been the most commonly used microorganisms for ethanol production. Yeast strains are generally chosen among *S. cerevisiae, S. ellypsoideuse, S. fragilis, S. carlsbergensis, Schizosaccharomyces pombe, Torula cremoris,* and *Candida pseudotropicalis.* Yeast species which can produce ethanol as the main fermentation product are reviewed, e.g., by Lin and Tanaka [8].

Among the ethanol-producing yeasts, the "industrial working horse" *S. cerevisiae* is by far the most well-known and most widely used yeast in industry and research for ethanol fermentation. This yeast can grow both on simple hexose sugars, such as glucose, and on the disaccharide sucrose. *S. cerevisiae* is also generally recognized to be safe as a food additive for human consumption and is therefore ideal for producing alcoholic beverages and for leavening bread. However, it cannot ferment pentoses such as xylose and arabinose to ethanol [14, 31]. There have been several research efforts to genetically modify *S. cerevisiae* to be able to consume xylose [33, 48–50]. Several attempts have been made to clone and express various bacterial genes, which is necessary for fermentation of xylose in *S. cerevisiae* [51, 52]. It resulted in great success, but probably not enough yet to efficiently ferment xylose with high yield and productivity [32].

Alternatively, xylose is converted to ethanol by some other naturally occurring recombinant. Among the wild-type xylose-fermenting yeast strains for ethanol production, *Pichia stipitis* and *C. shehatae* have reportedly shown promising results for industrial applications in terms of complete sugar utilization, minimal by-product formation, low sensitivity to temperature, and substrate concentration. Furthermore, *P. stipitis* is able to ferment a wide variety of sugars to ethanol and has no vitamin requirement for xylose fermentation [2].

Olsson and Hahn-Hägerdal [20] have presented a list of bacteria, yeasts, and filamentous fungi that produce ethanol from xylose. Certain species of the yeasts *Candida, Pichia, Kluyveromyces, Schizosaccharomyces,* and *Pachysolen* are among the naturally occurring organisms. Jeffries and Kurtzman [53] have reviewed the strain selection, taxonomy, and genetics of xylose-fermenting yeasts.

Utilization of cellobiose is important in ethanol production from lignocellulosic materials by SSF. However, a few ethanol-producing microorganisms are cellobiose-utilizing organisms. The requirement for addition of β-glucosidase has been eliminated by cellobiose utilization during fermentation, since presentation of cellobiose reduces the activity of cellulase. Cellobiose utilization eliminates the need for one class of cellulase enzymes [2]. *Brettanomyces custersii* is one of the yeasts identified as a promising glucose- and cellobiose-fermenting microorganism for SSF of cellulose for ethanol production [54].

High temperature tolerance could be a good characterization for ethanol production, since it simplifies fermentation cooling. On the other hand, one of the problems associated with SSF is the different optimum temperatures for saccharification and fermentation. Many attempts have been made to find thermotolerant yeasts for SSF. Szczodrak and Targonski [55] tested 58 yeast strains belonging to 12 different genera and capable of growing and fermenting sugars at temperatures of 40–46°C. They selected several strains belonging to the genera *Saccharomyces, Kluyveromyces,* and *Fabospora,* in view of their capacity to ferment glucose, galactose, and mannose at 40°C, 43°C, and 46°C, respectively. *Kluyveromyces marxianus* has been found to be a suitable strain for SSF [56].

3.11.2 Bacteria

A great number of bacteria are able to produce ethanol, although many of them generate multiple end products in addition to ethanol. *Zymomonas mobilis* is an unusual Gram-negative bacterium that has several appealing properties as a fermenting microorganism for ethanol production. It has a homoethanol fermentation pathway and tolerates up to 120 g/L ethanol. Its ethanol yield is comparable with *S. cerevisiae,* while it has much higher specific ethanol productivity (2.5×) than the yeast. However, the tolerance of *Z. mobilis* to ethanol is lower than that of *S. cerevisiae,* since some strains of *S. cerevisiae* can produce ethanol to give concentrations as high as 18% of the fermentation broth. The tolerance of *Z. mobilis* to inhibitors and low pH is also low. Similarly,

S. cerevisiae and *Z. mobilis* cannot utilize pentoses [14, 57]. Several genetic modifications have been performed for utilization of arabinose and xylose by *Z. mobilis*. However, *S. cerevisiae* has been more welcomed for industrial application, probably because of the industrial problems that may arise in working with bacteria. Separation of *S. cerevisiae* from fermentation media is much easier than separation of *Z. mobilis*, which is an important characteristic for reuse of the microorganisms in ethanol production processes.

Using genetically engineered bacteria for ethanol production is also applied in many studies. Ingram et al. [58] have reviewed metabolic engineering of bacteria for ethanol production. Recombinant *Escherichia coli* is a valuable bacterial resource for ethanol production. Construction of *E. coli* strains to selectively produce ethanol was one of the first successful applications of metabolic engineering. *E. coli* has several advantages as a biocatalyst for ethanol production, including the ability to ferment a wide spectrum of sugars, no requirements for complex growth factors, and prior industrial use (e.g., for production of recombinant protein). The major disadvantages associated with using *E. coli* cultures are a narrow and neutral pH growth range (6.0–8.0), less hardy cultures compared to yeast, and public perceptions regarding the danger of *E. coli* strains. Lack of data on the use of residual *E. coli* cell mass as an ingredient in animal feed is also an obstacle to its application [8].

Recently, the Japanese Research Institute of Innovative Technology for the Earth (RITE) developed a microorganism for ethanol production. The RITE strain is an engineered strain of *Corynebacterium glutamicum* that converts both pentose and hexose sugars into alcohol. The central metabolic pathway of *C. glutamicum* was engineered to produce ethanol. A recombinant strain that expressed the *Z. mobilis* gene coding for pyruvate decarboxylase and alcohol dehydrogenase was constructed [59]. RITE and Honda jointly developed a technology for production of ethanol production from lignocellulosic materials using the strain. It is claimed that application of this strain by using engineering technology from Honda enables a significant increase in alcohol conversion efficiency, in comparison to conventional cellulosic–bioethanol production processes.

3.11.3 Filamentous fungi

A great number of molds are able to produce ethanol. The filamentous fungi *Fusarium, Mucor, Monilia, Rhizopus, Ryzypose,* and *Paecilomyces* are among the fungi that can ferment pentoses to ethanol [33]. Zygomycetes are saprophytic filamentous fungi, which are able to produce several metabolites including ethanol. Among the three genera *Mucor, Rhizopus,* and *Rhizomucor, Mucor indicus* (formerly *M. rouxii*) and *Rhizopus oryzae* have shown good performances on ethanol productivity from glucose, xylose, and wood hydrolyzate [60]. *M. indicus* has several industrial advantages compared to baker's yeast for ethanol production, such as (a) capability of utilizing xylose, (b) having a valuable biomass, e.g., for production of chitosan, and (c) high optimum temperature of 37°C [61]. Skory et al. [62] examined 19 *Aspergilli* and 10 *Rhizopus* strains for their ability to ferment simple sugars (glucose, xylose, and arabinose) as well as complex substrates. An appreciable level of ethanol has been produced by *Aspergillus oryzae, R. oryzae,* and *R. javanicus.*

The dimorphic organism *M. circinelloides* is also used for production of ethanol from pentose and hexose sugars. Large amounts of ethanol have been produced during aerobic growth on glucose under nonoxygen-limiting conditions by this mold. However, ethanol production on galactose or xylose has been less significant [63]. Yields as high as 0.48 g/g ethanol from glucose by *M. indicus,* under anaerobic conditions, have been reported [64]. However, the yield and productivity of ethanol from xylose is lower than that of *P. stipitis* [65].

Although filamentous fungi have been industrially used for a long time for several purposes, a number of process-engineering problems are associated with these organisms due to their filamentous growth. Problems can appear in mixing, mass transfer, and heat transfer. Furthermore, attachment and growth on bioreactor walls, agitators, probes, and baffles cause heterogeneity within the bioreactor and problems in measurement of controlling parameters and cleaning of the bioreactor [66, 67]. Such potential problems might hinder industrial application of *M. indicus* for ethanol production. However, this fungus is dimorphic, and its morphology can be controlled to be yeast-like or pellet-like through fermentation [65].

3.12 Fermentation Process

In this section, we will discuss different fermentation processes applicable for ethanol production. Fermentation processes, as well as other biological processes, can be classified into batch, fed-batch, and continuous operation. All these methods are applicable in industrial fermentation of sugar substances and starch materials. These processes are well established, the fed-batch and continuous modes of operation being dominant in the ethanol market. When configuring the fermentation process, several parameters must be considered, including (a) high ethanol yield and productivity, (b) high conversion of sugars, and (c) low equipment cost. The need for detoxification and choice of the microorganism must be evaluated in relation to the fermentation configuration.

Presentation of a variety of inhibitors and their interaction effects, e.g., in lignocellulosic hydrolyzates, makes the fermentation process more complex than with other substrates for ethanol production [17, 21].

In fermentation of this hydrolyzate, the pentoses should be utilized in order to increase the overall yield of the process and to avoid problems in wastewater treatment. Therefore, it is still a challenge to use a hexose-fermenting organism such as S. cerevisiae for fermentation of the hydrolyzate.

When a mixture of hexoses and pentoses is present in the medium, microorganisms usually take up hexoses first and produce ethanol. As the hexose concentration decreases, they start to take up the pentose. Fermentation of hexoses can be successfully performed under anaerobic or microaerobic conditions, with high ethanol yield and productivity. However, fermentation of pentoses is generally a slow and aerobic process. If one adds air to ferment pentoses, the microorganisms will start utilizing the ethanol produced as well. It makes the entire process complicated and demands a well-designed and controlled process.

3.12.1 Batch processes

In batch processes, all nutrients required for fermentation are present in the medium prior to cultivation. Batch technology had been preferred in the past due to the ease of operation, low cost of controlling and moni-toring system, low requirements for complete sterilization, use of unskilled labor, low risk of financial loss, and easy management of feed-stocks. However, overall productivity of the process is very low, because of long turnaround times and an initial lag phase [9].

In order to improve traditional batch processes, cell recycling and application of several fermentors have been used. Reuse of produced cells can increase productivity of the process. Application of several fermentors operated at staggered intervals can provide a continuous feed to the distillation system. One of the successful batch methods applied for industrial production of ethanol is Melle-Boinot fermen-tation. This process achieves a reduced fermentation time and increased yield by recycling yeast and applying several fermentors operated at staggered intervals. In this method, yeast cells from pre-vious fermentation are separated from the media by centrifugation and up to 80% are recycled [9, 68]. Instead of centrifugation, the cells can be filtered, followed by the separation of yeast from the filter aid using hydrocyclones and then recycled [69].

In well-detoxified or completely noninhibiting acid hydrolyzates of lignocellulosic materials, exponential growth will be obtained after inoc-ulation of the bioreactor. If the hydrolyzate is slightly inhibiting, there will be a relatively long lag phase during which part of the inhibitors are converted. However, if the hydrolyzate is severely inhibiting, no conversion of the inhibitors will occur, and neither cell growth nor fer-mentation will occur. A slightly inhibiting hydrolyzate can thus be detoxified during batch fermentation. However, very high concentration of the inhibitors will cause complete inactivation of the metabolism [18].

Several strategies may be considered for fermentation of hydrolyzate to improve the in situ detoxification in batch fermentation and obtain higher yield and productivity of ethanol. Having high initial cell density, increasing the tolerance of microorganisms against the inhibitors by either adaptation of cells to the medium or genetic modification of the microorganism, and choosing optimal reactor conditions to minimize the effects of inhibitors are among these strategies.

Volumetric ethanol productivity is low in lignocellulosic hydrolyzates when low cell-mass inocula are used due to poor cell growth. Usually, high cell concentration, e.g., 10 g/L dry cells, have been used in order to find a high yield and productivity of ethanol in different studies. In addition, a high initial cell density helps the process for in situ detoxification by the microorganisms, and therefore, the demand for a detoxification unit decreases. In situ detoxification of the inhibitors may even lead to increased ethanol yield and productivity, due to uncoupling by the presence of weak acids, or due to decreased glycerol production in the presence of furfural [21]. Adaptation of the cells to hydrolyzate or genetic modification of the microorganism can significantly improve the yield and productivity of ethanol. Optimization of reactor conditions can be used to minimize the effects of inhibitors. Among the different parameters, cell growth is found to be strongly dependent on pH [18, 21].

3.12.2 Fed-batch processes

In fed-batch processes (or semi-continuous processes), the substrate and required nutrients are added continuously or intermittently to the initial medium after the start of cultivation or from the point halfway through the batch process. Fed-batch processes have been utilized to avoid utilizing substrates that inhibit growth rate if present at high concentration, to overcome catabolic repression, to demand less initial biomass, to overcome the problem of contamination, and to avoid mutation and plasmid instability found in continuous culture. Furthermore, fed-batch processes do not face the problem of washout, which can occur in continuous fermentation. A major disadvantage of a fed-batch process is the need for additional control instruments that require a substantial amount of operator skill. In addition, for systems without feedback control, where the feed is added on a predetermined fixed schedule, there can be difficulty in dealing with any deviation (i.e., time courses may not always follow the expected profiles) [70]. The fed-batch processes without feedback control can be classified as intermittent fed-batch, constant-rate fed-batch, exponential fed-batch, and optimized fed-batch.

The fed-batch processes with feedback control have been classified as indirect-control and direct-control fed-batch processes [70, 71].

The fed-batch technique is one of the promising methods for fermentation of dilute-acid hydrolyzates of lignocellulosic materials. The basic concept behind the success of this technique is the capability of in situ detoxification of hydrolyzates by the fermenting microorganisms. Since the yeast has a limited capacity for conversion of the inhibitors, the achievement of a successful fermentation strongly depends on the feed rate of the hydrolyzate. By adding the substrate at a low rate in fed-batch fermentation, the concentrations of bioconvertible inhibitors such as furfural and HMF in the fermentor remain low, and the inhibiting effect therefore decreases. At a very high feed rate, using an inhibiting hydrolyzate, both ethanol production and cell growth can stop, whereas at a very low feed rate, the hydrolyzate may still be converted, but at a very low productivity rate, which has been experimentally confirmed. Consequently, there should exist an optimum feed rate [15, 18, 21].

Similar to batch operations, higher optimum dilution rate in fed-batch cultivation can be obtained by (a) high initial cell concentration, (b) increasing the tolerance of microorganisms against the inhibitors, and (c) choosing optimal reactor conditions to minimize the effects of inhibitors. Productivity in fed-batch fermentation is generally limited by the feed rate which, in turn, is limited by the cell-mass concentration [21].

3.12.3   Continuous processes

Process design studies of molasses fermentation have shown that the investment cost was considerably reduced when continuous rather than batch fermentation was employed, and that the productivity of ethanol could be increased by more than 200%. Continuous operations can be classified into continuous fermentation with or without feedback control. In continuous fermentation without feedback control, called a *chemostat*, the feed medium containing all the nutrients is continuously fed at a constant rate (dilution rate D) and the cultured broth is simultaneously removed from the fermentor at the same rate. The chemostat is quite useful in the optimization of media formulation and to investigate the physiological state of the microorganism [71]. Continuous fermentations with feedback control are turbidostat, phauxostat, and nutristat. A turbidostat with feedback control is a continuous process to maintain the cell concentration at a constant level by controlling the medium feeding rate. A phauxostat is an extended nutristat, which maintains the pH value of the medium in the fermentor at a preset value. A nutristat with feedback control is a cultivation technique to maintain nutrient concentration at a constant level [71].

When lignocellulosic hydrolyzates are added at a low feed rate in continuous fermentation, low concentration of bioconvertible inhibitors in the fermentor is assured. In spite of a number of potential advantages in terms of productivity, this method has not developed much yet in fermentation of the acid hydrolyzates. One should consider the following points in continuous cultivation of acid hydrolyzates of lignocelluloses:

■ Cell growth is necessary at a rate equal to the dilution rate in order to avoid washout of the cells in continuous cultivation.

■ Growth rate is low in fermentation of hydrolyzates because of the presence of inhibitors.

■ The cells should keep their viability and vitality for a long time.

The major drawback of the continuous fermentation is that, in contrast to the situation in fed-batch fermentation, cell growth is necessary at a rate equal to the dilution rate, in order to avoid washout of the cells in continuous cultivation [21]. The productivity is a function of the dilution rate, and since the growth rate is decreased by the inhibitors, the productivity in continuous fermentation of lignocellulosic hydrolyzates is low. Furthermore, at a very low dilution rate, the conversion rate of the inhibitors can be expected to decrease due to the decreased specific growth rate of the biomass. Thus, washout may occur even at very low dilution rates [18]. On the other hand, one of the major advantages of continuous cultivation is the possibility to run the process for a long time (e.g., several months), whereas the microorganisms usually lose their activity after facing the inhibitory conditions of the hydrolyzate. By employing cell-retention systems, the cell-mass concentration in the fermentor, the maximum dilution rate, and thus the maximum ethanol productivity increase. Different cell-retention systems have been investigated by cell immobilization and encapsulation, and cell recirculation by filtration, settling, and centrifugation. A relatively old study [72] shows that the investment cost for a continuous process with cell recirculation has been found to be less than that for continuous fermentation without cell recirculation.

Biostil® is the trade name of a continuous industrial process for ethanol production with partial recirculation of both yeast and wastewater. The fermentor works continuously; the cells are separated by using a centrifuge, and a part of the separated cells is returned to the fermentor. Most of the ethanol-depleted beer including residual sugars is then recycled to the fermentor. In this process, besides providing enough cell concentration in the fermentor, less water is consumed and a more concentrated stillage is produced. Therefore, the process has a lower wastewater problem. However, the process needs a special type of centrifuge (which is expensive) in order to avoid deactivation of the cells [47, 73].

Application of an encapsulated cell system in continuous cultivation has several advantages, compared to either a free-cell or traditionally entrapped cell system, e.g., in alginate matrix. Encapsulation provides higher cell concentrations than free-cell systems in the medium, which leads to higher productivity per volume of the bioreactor in continuous cultivation. Furthermore, the biomass can easily be separated from the medium without centrifugation or filtration. The advantages of encapsulation, compared to cell entrapment, are less resistance to diffusion through beads/capsules, some degree of freedom in movement of the encapsulated cells, no cell leakage from the capsules, and higher cell concentration [74].

3.12.4   Series-arranged continuous flow fermentation

Ethanol can be produced by using continuous flow fermentors arranged in a series with complete sugar utilization or high ethanol concentration. With two fermentors arranged in a series, the retention time can be chosen so that the sugar is only partially utilized in the first, with fermentation completed in the second. Ethanol inhibition is reduced in the first fermentor, allowing a faster throughput. The second, lower-productivity fermentor can now convert less sugar than if operated alone. For high product concentration, productivity of a two-stage system has been 2.3 times higher than that of a single stage [47, 75].

A two-stage continuous ethanol fermentation process with yeast recirculation is used industrially by Danish Distilleries Ltd., Grena, for molasses fermentation (see Fig. 3.8). Two fermentors with 170,000-L volume produce 66 g/L ethanol in 21-h retention time [76].

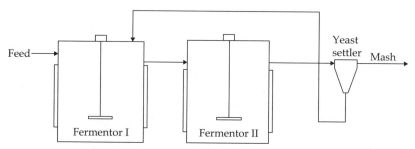

Figure 3.8 Two-stage continuous ethanol fermentation process with yeast recirculation [76, 77]. (A seven-fermentor-series system (70,000-L volume each fermentor) was also used in the Netherlands to produce 86 g/L ethanol in 8-h retention time [78]. A Japanese company used a six-fermentor-series system (total volume 100,000 L) with 8.5-h retention time to produce 95 g/L ethanol [79].)

3.12.5 Strategies for fermentation of enzymatic lignocellulosic hydrolyzates

The cellulose fraction of lignocelluloses can be converted to ethanol by either simultaneous saccharification and fermentation (SSF) or separate enzymatic hydrolysis and fermentation (SHF) processes. A schematic of these processes is shown in Fig. 3.9. It is also possible to combine the cellulase production, enzymatic hydrolysis, and fermentation in one step, called direct microbial conversion (DMC). There are cost savings because of the reduced number of required vessels. However, there is less attention to DMC for industrial purposes because of the low ethanol yield in DMC, formation of several by-products, and low ethanol tolerance of the microorganisms used [2].

3.12.6 Separate enzymatic hydrolysis and fermentation (SHF)

In SHF, enzymatic hydrolysis for conversion of pretreated cellulose to glucose is the first step. Produced glucose is then converted to ethanol in the second step. Enzymatic hydrolysis can be performed in the optimum conditions of the cellulase. The optimum temperature for hydrolysis by cellulase is usually between 45°C and 50°C, depending on the microorganism that produces the cellulase. The major disadvantage of SHF is that the released sugars severely inhibit cellulase activity. The activity of cellulose is reduced by 60% at a cellobiose concentration as

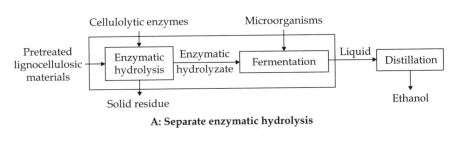

A: Separate enzymatic hydrolysis

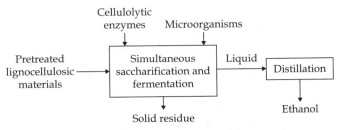

B: Simultaneous saccharification and fermentation

Figure 3.9 Main steps in SSF or SHF for ethanol production.

low as 6 g/L. Although glucose also decreases the cellulase activity, the inhibitory effect of glucose is lower than that of cellobiose. On the other hand, glucose is a strong inhibitor for β-glucosidase. At a level of 3 g/L of glucose, the activity of β-glucosidase reduces by 75% [27, 80]. Another possible problem in SHF is contamination. Hydrolysis is a lengthy process (one or possibly several days), and a dilute solution of sugar always has a risk of contamination, even at rather high temperatures such as 45–50°C.

3.12.7   Simultaneous saccharification and fermentation (SSF)

SSF combines enzymatic hydrolysis of cellulose and fermentation in one step. As cellulose converts to glucose, a fermenting microorganism is presented in the medium and it immediately consumes the glucose produced. As mentioned, cellobiose and glucose significantly decrease the activity of cellulase. SSF gives higher reported ethanol yields and requires lower amounts of enzyme, because end-product inhibition from cellobiose and glucose formed during enzymatic hydrolysis is relieved by the yeast fermentation. SSF has the following advantages compared to SHF:

- Fewer vessels are required for SSF, in comparison to SHF, resulting in capital cost savings.

- Less contamination during enzymatic hydrolysis, since the presence of ethanol reduces the possibility of contamination.

- Higher yield of ethanol.

- Lower enzyme-loading requirement.

On the other hand, SSF has the following drawbacks compared to SHF:

- SSF requires that enzyme and culture conditions be compatible with respect to pH and temperature. In particular, the difference between optimum temperatures of the hydrolyzing enzymes and fermenting microorganisms is usually problematic. *Trichoderma reesei* cellulases, which constitute the most active preparations, have optimal activity between 45°C and 50°C, whereas *S. cerevisiae* has an optimum temperature between 30°C and 35°C. The optimal temperature for SSF is around 38°C, which is a compromise between the optimal temperatures for hydrolysis and fermentation. Hydrolysis is usually the rate-limiting process in SSF [27]. Several thermotolerant yeasts (e.g., *C. acidothermophilum* and *K. marxianus*) and bacteria have been used in SSF to raise the temperature close to the optimal hydrolysis temperature.

■ Cellulase is inhibited by ethanol. For instance, at 30 g/L ethanol, the enzyme activity was reduced by 25% [2]. Ethanol inhibition may be a limiting factor in production of high ethanol concentration. However, there has been less attention to ethanol inhibition of cellulase, since practically it is not possible to work with very high substrate concentration in SSF, because of the problem with mechanical mixing.

■ Another problem arises from the fact that most microorganisms used for converting cellulosic feedstock cannot utilize xylose, a hemicellulose hydrolysis product [8].

3.12.8 Comparison between enzymatic and acid hydrolysis for lignocellulosic materials

The two most promising processes for industrial production of ethanol from cellulosic materials are two-stage dilute-acid hydrolysis (a chemical process) and SSF (an enzymatic process). Advantages and disadvantages of dilute-acid and enzymatic hydrolyses are summarized in Table 3.3. Enzymatic hydrolysis is carried out under mild conditions, whereas high temperature and low pH result in corrosive conditions for acid hydrolysis. While it is possible to obtain a cellulose hydrolysis of close to 100% by enzymatic hydrolysis after a pretreatment, it is difficult to achieve such a high yield with acid hydrolysis. The yield of conversion of cellulose to sugar with dilute-acid hydrolysis is usually less than 60%. Furthermore, the previously mentioned inhibitory compounds are formed during acid hydrolysis, whereas this problem is not so severe for enzymatic hydrolysis. Acid hydrolysis conditions may destroy nutrients sensitive to acid and high temperature such as vitamins, which may introduce the process together with the lignocellulosic materials.

TABLE 3.3 Advantages and Disadvantages of Dilute-Acid and Enzymatic Hydrolyses

Parameters	Dilute-acid hydrolysis	Enzymatic hydrolysis
Rate of hydrolysis	Very high	Low
Overall yield of sugars	Low	High and depend upon pretreatment
Catalyst costs	Low	High
Conditions	Harsh reaction conditions (e.g., high pressure and temperature)	Mild conditions (e.g., 50°C, atmospheric pressure, pH 4.8)
Inhibitors formation	Highly inhibitory hydrolyzate	Noninhibitory hydrolyzate
Degradation of sensitive nutrients such as vitamins	High	Low

On the other hand, enzymatic hydrolysis has its own problems in comparison to dilute-acid hydrolysis. Hydrolysis for several days is necessary for enzymatic hydrolysis, whereas a few minutes are enough for acid hydrolysis. The prices of the enzymes are still very high, although a new development has claimed a 30-fold decrease in the price of cellulase.

3.13 Ethanol Recovery

Fermented broth or "mash" typically contains 2–12% ethanol. Furthermore, it contains a number of other materials that can be classified into microbial biomass, fusel oil, volatile components, and stillage. Fusel oil is a mixture of primary methylbutanols and methylpropanols formed from α-ketoacids and derived from or leading to amino acids. Depending on the resources used, important components of fusel oil can be isoamylalcohol, n-propylalcohol, sec-butylalcohol, isobutylalcohol, n-butlyalcohol, active amylalcohol, and n-amylalcohol. The amount of fusel oil in mash depends on the pH of the fermentor. Fusel oil is used in solvents for paints, polymers, varnishes, and essential oils. Acetaldehyde and trace amounts of other aldehydes and volatile esters are usually produced from grains and molasses. Typically, 1 L of acetaldehyde and 1–5 L of fusel oil are produced per 1000 L of ethanol [9, 47].

Stillage consists of the nonvolatile fraction of materials remaining after alcohol distillation. Its composition depends greatly on the type of feedstock used for fermentation. Stillage generally contains solids, residual sugars, residual ethanol, waxes, fats, fibers, and mineral salts. The solids may be originated from feedstock proteins and spent microbial cells [9].

3.14 Distillation

Mash is usually centrifuged or settled in order to separate the microbial biomass from the liquid and then sent to the ethanol recovery system. Distillation is typically used for the separation of ethanol, aldehydes, fusel oil, and stillage [9]. Ethanol is readily concentrated from mash by distillation, since the volatility of ethanol in a dilute solution is much higher than the volatility of water. Therefore, ethanol is separated from the rest of the materials and water by distillation. However, ethanol and water form an azeotrope at 95.57 wt% ethanol (89 mol% ethanol) with a minimum boiling point of 78.15°C. This mixture behaves as a single component in a simple distillation, and no further enrichment than 95.57 wt% of ethanol can be achieved by simple distillation [9, 47, 81]. Various industrial distillation systems for ethanol purification are (a) simple two-column systems, (b) three- or four-column barbet systems, (c) three-column Othmer system, (d) vacuum rectification,

(e) vapor recompression, (f) multieffect distillation, and (g) six-column reagent alcohol system [9, 47]. These methods are reviewed by Kosaric [9]. The following parameters should be considered for selection of the industrial distillation systems:

- Energy consumption (e.g., steam consumption or cooling water consumption per kilogram of ethanol produced).

- Quality of ethanol (complete separation of fusel oil and light components).

- How to deal with the problem associated with clogging of the first distillation column and its reboiler because of precipitation or formation of solids. Special design and use of a vacuum may be applied for overcoming the problem in the column. Using open steam instead of application of a reboiler can prevent clogging of the reboiler, in spite of the increase in amount of wastewater.

- Simplicity in controlling the system.

- Simplicity in opening column parts and cleaning the columns.

Of course, lower capital investment is also one of the main parameters in the selection of distillation systems.

Ethanol is present in the market with different degrees of purity. The majority of ethanol is 190 proof (95% or 92.4%, minimum) used for solvent, pharmaceutical, cosmetic, and chemical applications. Technical-grade ethanol, containing up to 5% volatile organic aldehyde, ester, and sometimes methanol, is used for industrial solvents and chemical syntheses. High-purity 200 proof (99.85%) anhydrous ethanol is produced for special chemical applications. For fuel use in mixture with gasoline (gasohol), nearly anhydrous (99.2%) ethanol, but with higher available levels of organic impurities, is used [47].

A simple two-column system is described here, while other systems are presented in the literature (e.g., [9, 47]). Simple one- or two-column systems with only a stripping and rectification section are usually used to produce lower-quality industrial alcohol and azeotrope alcohol for further dehydration to fuel grade. The simplest continuous ethanol distillation system consists of stripping and rectification sections, either together in one column or separated into two columns (see Fig. 3.10).

The mash produced is pumped into a continuous distillation process, where steam is used to heat the mash to its boiling point in the stripper column. The ethanol-enriched vapors pass through a rectifying column and are condensed and removed from the top of the rectifier at around 95% ethanol. The ethanol-stripped stillage falls to the bottom of the stripper column and is pumped to a stillage tank. Aldehydes are drawn

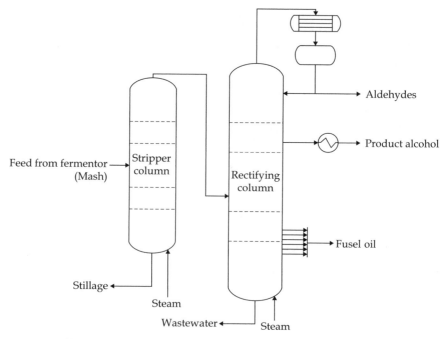

Figure 3.10 Two-column system for distillation of ethanol.

from the head vapor, condensed, and partly used as reflux. Fusel oil is taken out from several plates of the rectifying section [9, 47, 82].

With efficient distillation, the stillage should contain less than 0.1% ethanol since the presence of ethanol significantly increases the chemical oxygen demand (COD) of wastewater. For each 1% ethanol left in the stillage, the COD of the stillage is incremented by more than 20 g/L. Due to the potential impact of residual ethanol content, therefore, proper control over distillation can greatly affect the COD of stillage [82].

3.15 Alternative Processes for Ethanol Recovery and Purification

Since distillation is a highly energy-consuming process, several processes have been developed for purification of ethanol from fermentation broth: for example, solvent extraction, CO_2 extraction, vapor recompression systems, and low-temperature blending with gasoline [9]. However, these processes are not established in the industrial production of ethanol.

3.16 Ethanol Dehydration

In order to allow blending of alcohol with gasoline, the water content of ethanol must be reduced to less than 1% by volume, which is not possible by distillation. Higher water levels can result in phase separation of an alcohol–water mixture from the gasoline phase, which may cause engine malfunction. Removal of water beyond the last 5% is called *dehydration* or *drying of ethanol*. Azeotropic distillation was previously employed to produce higher-purity ethanol by adding a third component, such as benzene, cyclohexane, or ether, to break the azeotrope and produce dry ethanol [82]. To avoid illegal transfer of ethanol from the industrial market into the potable alcohol market, where it is highly regulated and taxed, dry alcohol usually requires the addition of denaturing agents that render it toxic for human consumption; the azeotropic reagents conveniently meet this requirement [82]. Except in the high-purity reagent-grade ethanol market, azeotropic drying has been supplanted by molecular sieve drying technology.

3.16.1 Molecular sieve adsorption

The molecular sieve is a more energy-efficient method than azeotropic distillation. Furthermore, this method avoids the occupational hazards associated with azeotropic chemical admixtures. In molecular sieve drying, 95% ethanol is passed through a bed of synthetic zeolite with uniform pore sizes that preferentially adsorb water molecules. Approximately three-fourths of adsorbed material is water and one-fourth is ethanol. The bed becomes saturated after a few minutes and must be regenerated by heating or evacuation to drive out the adsorbed water. During the regeneration phase, a side stream of ethanol/water (often around 50%) is produced, which must be redistilled before returning to the drying process [82].

3.16.2 Membrane technology

Membranes can also be used for ethanol purification. Reverse osmosis (RO), which employs membranes impermeable to ethanol and permeable to water, can be used for purification of ethanol from water. Using a membrane permeable to ethanol but not to water is another approach [9]. Pervaporation, a promising membrane technique for separation of organic liquid mixtures such as azeotropic mixtures or near-boiling-point mixtures, can also be used for separation of these azeotropes [81, 83]. It involves the separation of ethanol–water azeotrope or near-azeotropic ethanol–water composition (from about 95 to 99.5 wt% ethanol) through water-permeable (or water-selective) membranes to remove the rest of the water from the concentrated ethanol solution [84].

3.17   Concluding Remarks and Future Prospects

Ethanol has been well established in the fuel market, where its share from less than 1 GL in 1975 is expected to increase to 100 GL in 2015. Grains, sugarcane juice, and molasses are the dominant raw materials for the time being, while lignocellulosic materials are expected to have a significant share in this market in the future. There have been great achievements in the development of ethanol production from lignocellulosic materials, and large-scale facilities are expected to be built within a few years. However, several challenges will still persist in this process in the future, until the process is fully established.

References

1. H. Rothman, R. Greenshields, and F. R. Calle. *The Alcohol Economy: Fuel Ethanol and the Brazilian Experience,* First ed., London: Francis Printer, 1983.
2. C. E. Wyman. *Handbook on Bioethanol: Production and Utilization*, Washington, DC: Taylor & Francis, 1996.
3. C. Breidenich, D. Magraw, A. Rowley, and J. Rubin, The Kyoto protocol to the United Nations framework convention on climate change, *American Journal of International Law* **92**, 315–331, 1998.
4. F. O. Licht. World Ethanol Markets: The Outlook to 2015, Tunbridge Wells, UK: F. O. Licht, 2006.
5. C. N. Hamelinck and A. P. C. Faaij. Outlook for advanced biofuels, *Energy Policy* **34**, 3268–3283, 2006.
6. J. M. Marchetti, V. U. Miguel, and A. F. Errazu. Possible methods for biodiesel production, *Renewable & Sustainable Energy Reviews*, **11**, 1300–1311, 2006.
7. M. Roehr. *The Biotechnology of Ethanol: Classical and Future Applications*, First ed., Weinheim: Wiley-VCH, 2001.
8. Y. Lin and S. Tanaka. Ethanol fermentation from biomass resources: Current state and prospects, *Applied Microbiology and Biotechnology* **69**, 627–642, 2006.
9. N. Kosaric, A. Wieczorirek, G. P. Cosentono, and R. J. Magee. Ethanol fermentation, In: *Biotechnology: A comprehensive Treatise*, Reed, G. (Ed.), Weinheim, Germany. Verlag-Chemie, 1983.
10. M. Gulati, K. Kohlmann, M. R. Ladisch, R. Hespell, and R. J. Bothast. Assessment of ethanol production options for corn products, *Bioresource Technology* **58**, 253–264, 1996.
11. J. E. Bailey and D. F. Ollis, *Biochemical Engineering Fundamentals*, Second ed., Singapore: McGraw-Hill, 1986.
12. Y. Sun and J. Cheng. Hydrolysis of lignocellulosic materials for ethanol production: A review, *Bioresource Technology* **83**, 1–11, 2002.
13. K. Karimi, S. Kheradmandinia, and M. J. Taherzadeh. Conversion of rice straw to sugars by dilute-acid hydrolysis, *Biomass and Bioenergy* **30**, 247–253, 2006.
14. B. C. Saha. Hemicellulose bioconversion, *Journal of Industrial Microbiology and Biotechnology* **30**, 279–291, 2003.
15. M. J. Taherzadeh and K. Karimi. Acid-based hydrolysis processes for ethanol from lignocellulosic materials: A review, *Bioresources*, **2**, 472-499, 2007.
16. P. Béguin and J.-P. Aubert. The biological degradation of cellulose, *FEMS Microbiology Reviews* **13**, 25–58, 1994.
17. E. Palmqvist and B. Hahn-Hägerdal. Fermentation of lignocellulosic hydrolysates. II: Inhibitors and mechanisms of inhibition, *Bioresource Technology* **74**, 25–33, 2000.
18. M. J. Taherzadeh and C. Niklasson. Ethanol from lignocellulosic materials: Pretreatment, acid and enzymatic hydrolyses and fermentation, In: *Lignocellulose Biodegradation*, Saha B. C. and Hayashi, K. (Eds.), First ed., Washington, DC: American Chemical Society, 2004.

19. M. J. Taherzadeh, C. Niklasson, and G. Lidén. Acetic acid—Friend or foe in anaerobic batch conversion of glucose to ethanol by *Saccharomyces cerevisiae*? *Chemical Engineering Science* **52**, 2653–2659, 1997.
20. L. Olsson and B. Hahn-Hägerdal. Fermentation of lignocellulosic hydrolysates for ethanol production, *Enzyme and Microbial Technology* **18**, 312–331, 1996.
21. E. Palmqvist and B. Hahn-Hägerdal. Fermentation of lignocellulosic hydrolysates. I: Inhibition and detoxification, *Bioresource Technology* **74**, 17–24, 2000.
22. P. Persson, J. Andersson, L. Gorton, S. Larsson, N. O. Nilvebrant, and L. J. Jonsson. Effect of different forms of alkali treatment on specific fermentation inhibitors and on the fermentability of lignocellulose hydrolysates for production of fuel ethanol, *Journal of Agricultural and Food Chemistry* **50**, 5318–5325, 2002.
23. R. Millati, C. Niklasson, and M. J. Taherzadeh. Effect of pH, time and temperature of overliming on detoxification of dilute-acid hydrolyzates for fermentation by *Saccharomyces cerevisiae*, *Process Biochemistry* **38**, 515–522, 2002.
24. A. Martinez, M. E. Rodriguez, M. L. Wells, S. W. York, J. F. Preston, and L. O. Ingram. Detoxification of dilute acid hydrolysates of lignocellulose with lime, *Biotechnology Progress* **17**, 287–293, 2001.
25. S. Amartey and T. Jeffries. An improvement in *Pichia stipitis* fermentation of acid-hydrolysed hemicellulose achieved by overliming (calcium hydroxide treatment) and strain adaptation, *World Journal of Microbiology & Biotechnology* **12**, 281–283, 1996.
26. R. Purwadi, C. Niklasson, and M. J. Taherzadeh. Kinetic study of detoxification of dilute-acid hydrolyzates by $Ca(OH)_2$, *Journal of Biotechnology* **114**, 187–198, 2004.
27. G. P. Philippidis and T. K. Smith. Limiting factors in the simultaneous saccharification and fermentation process for conversion of cellulosic biomass to fuel ethanol, *Applied Biochemistry and Biotechnology*, **51/52**, 117–124, 1995.
28. K. Stenberg, M. Bollok, K. Reczey, M. Galbe, and G. Zacchi. Effect of substrate and cellulase concentration on simultaneous saccharification and fermentation of steam-pretreated softwood for ethanol production, *Biotechnology and Bioengineering* **68**, 204–210, 2000.
29. M. Linde, M. Galbe, and G. Zacchi. Simultaneous saccharification and fermentation of steam-pretreated barley straw at low enzyme loadings and low yeast concentration, *Enzyme and Microbial Technology* **40**, 1100–1107, 2007.
30. M. Alkasrawi, T. Eriksson, J. Borjesson, A. Wingren, M. Galbe, F. Tjerneld, G. Zacchi. The effect of Tween-20 on simultaneous saccharification and fermentation of softwood to ethanol, *Enzyme and Microbial Technology* **33**, 71–78, 2003.
31. T. Brandberg. Fermentation of undetoxified dilute acid lignocellulose hydrolyzate for fuel ethanol production, *Chemical Reaction Engineering,* Chalmers University of Technology, Göteborg, Sweden, 2005.
32. M. Sonderegger, M. Jeppsson, C. Larsson, M. F. Gorwa-Grauslund, E. Boles, L. Olsson et al. Fermentation performance of engineered and evolved xylose-fermenting *Saccharomyces cerevisiae* strains, *Biotechnology and Bioengineering* **87**, 90–98, 2004.
33. T. W. Jeffries. Engineering yeasts for xylose metabolism, *Current Opinion in Biotechnology* **17**, 320–326, 2006.
34. A. J. van Maris, D. A. Abbott, E. Bellissimi, J. van den Brink, M. Kuyper, M. A. Luttik et al. Alcoholic fermentation of carbon sources in biomass hydrolysates by *Saccharomyces cerevisiae*: Current status, *Antonie Van Leeuwenhoek* **90**, 391–418, 2006.
35. M. Jeppsson, B. Johansson, B. Hahn-Hägerdal, and M. F. Gorwa-Grauslund. Reduced oxidative pentose phosphate pathway flux in recombinant xylose-utilizing *Saccharomyces cerevisiae* strains improves the ethanol yield from xylose, *Applied and Environmental Microbiology* **68**, 1604–1609, 2002.
36. M. Desvaux, E. Guedon, and H. Petitdemange. Cellulose catabolism by *Clostridium cellulolyticum* growing in batch culture on defined medium, *Applied and Environmental Microbiology*, **66**, 2461–2470, 2000.
37. I. S. Horváth, M. J. Taherzadeh, C. Niklasson, and G. Lidén. Effects of furfural on anaerobic continuous cultivation of *Saccharomyces cerevisiae*, *Biotechnology and Bioengineering* **75**, 540–549, 2001.

38. W. H. Kampen, Todaro CL. Nutritional requirements in fermentation processes, In: *Fermentation and Biochemical Engineering Handbook*, Second ed., Noyes Publications, 1997.
39. M. J. Taherzadeh, L. Adler, and G. Lidén. Strategies for enhancing fermentative production of glycerol—A review, *Enzyme and Microbial Technology* 31, 53–66, 2002.
40. R. Millati. Ethanol production from lignocellulosic materials, *Chemical Reaction Engineering*, Chalmers University of Technology, Göteborg, Sweden, 2005.
41. H. Schneider. Conversion of pentoses to ethanol by yeast and fungi, *Critical Reviews in Biotechnology* 9, 1–40 1989.
42. K. Karhumaa, R. Fromanger, B. Hahn-Hägerdal, and M. F. Gorwa-Grauslund. High activity of xylose reductase and xylitol dehydrogenase improves xylose fermentation by recombinant *Saccharomyces cerevisiae*, *Applied Microbiology and Biotechnology* 73, 1039–1046, 2006.
43. M. Jeppsson, K. Traff, B. Johansson, B. Hahn-Hagerdal, and M. F. Gorwa-Grauslund. Effect of enhanced xylose reductase activity on xylose consumption and product distribution in xylose-fermenting recombinant *Saccharomyces cerevisiae*, *FEMS Yeast Research* 3, 167–175, 2003.
44. Y. C. Bor, C. Moraes, S. P. Lee, W. L. Crosby, A. J. Sinskey, and C. A. Batt. Cloning and sequencing the *Lactobacillus brevis* gene encoding xylose isomerase, *Gene* 114, 127–132, 1992.
45. X. Zhu, M. Teng, L. Niu, C. Xu, and Y. Wang. Structure of xylose isomerase from *Streptomyces diastaticus* No. 7 strain M1033 at 1.85 Å resolution, *Acta Crystallographica. Section D, Biological Crystallography* 56, 129–136, 2000.
46. Y. Wang, Z. Huang, X. Dai, J. Liu, T. Cui, L. Niu, C. Wang, and X. Xu. The sequence of xylose isomerase gene from *Streptomyces diastaticus* No. 7 M1033, *Clinical Journal of Biotechnology* 10, 97–103, 1994.
47. B. L. Maiorella. Ethanol, In: *Comprehensive Biotechnology*, Moo-Young, M. (Ed.), First ed., Oxford: Pergamon Press Ltd., 1985.
48. S. Govindaswamy and L. M. Vane. Kinetics of growth and ethanol production on different carbon substrates using genetically engineered xylose-fermenting yeast, *Bioresource Technology* 98, 677–685, 2007.
49. C. Martin, M. Marcet, O. Almazan, and L. J. Jonsson. Adaptation of a recombinant xylose-utilizing *Saccharomyces cerevisiae* strain to a sugarcane bagasse hydrolysate with high content of fermentation inhibitors, *Bioresource Technology* 98, 1767–1773, 2006.
50. L. Ruohonen, A. Aristidou, A. D. Frey, M. Penttila, and P. T. Kallio. Expression of *Vitreoscilla* hemoglobin improves the metabolism of xylose in recombinant yeast *Saccharomyces cerevisiae* under low oxygen conditions, *Enzyme and Microbial Technology* 39, 6–14, 2006.
51. N. Ho, Z. Chen, and A. Brainard. Genetically engineered *Saccharomyces* yeast capable of effective cofermentation of glucose and xylose, *Applied Environmental Microbiology* 64, 1852–1859, 1998.
52. M. H. Toivari, A. Aristidou, L. Ruohonen, and M. Penttila. Conversion of xylose to ethanol by recombinant *Saccharomyces cerevisiae*: Importance of xylulokinase (XKS1) and oxygen availability, *Metabolic Engineering* 3, 236–249, 2001.
53. T. W. Jeffries and C. P. Kurtzman. Strain selection, taxonomy, and genetics of xylose-fermenting yeasts, *Enzyme and Microbial Technology* 16, 922–932, 1994.
54. D. D. Spindler, C. E. Wyman, K. Grohmann, and G. P. Philippidis. Evaluation of the cellobiose-fermenting yeast *Brettanomyces custersii* in the simultaneous saccharification and fermentation of cellulose, *Biotechnology Letters* 14, 403–407, 1992.
55. J. Szczodrak and Z. Targonski. Selection of thermotolerant yeast strains for simultaneous saccharification and fermentation of cellulose, *Biotechnology and Bioengineering* 31, 300–303, 1988.
56. M. Ballesteros, J. M. Oliva, M. J. Negro, P. Manzanares, and I. Ballesteros. Ethanol from lignocellulosic materials by a simultaneous saccharification and fermentation process (SSF) with *Kluyveromyces marxianus* CECT 10875, *Process Biochemistry* 39, 1843–1848, 2004.
57. I. S. Horvath. Fermentation inhibitors in the production of bio-ethanol, *Chemical Reaction Engineering*, Chalmers University of Technology, Göteborg, Sweden, 2004.

58. L. Ingram, P. Gomez, X. Lai, M. Monirruzzamam, B. Wood, L. Yomano, and S. York. Metabolic engineering of bacteria for ethanol production, *Biotechnology and Bioengineering* **58**, 204–214, 1998.
59. M. Inui, H. Kawaguchi, S. Murakami, A. A. Vertes, and H. Yukawa. Metabolic engineering of *Corynebacterium glutamicum* for fuel ethanol production under oxygen-deprivation conditions, *Journal of Molecular Microbiology and Biotechnology* **8**, 243–254, 2004.
60. R. Millati, L. Edebo, and M. J. Taherzadeh. Performance of *Rhizopus, Rhizomucor,* and *Mucor* in ethanol production from glucose, xylose, and wood hydrolyzates, *Enzyme and Microbial Technology* **36**, 294–300, 2005.
61. K. Karimi, G. Emtiazi, and M. J. Taherzadeh. Production of ethanol and mycelial biomass from rice straw hemicellulose hydrolyzate by *Mucor indicus*, *Process Biochemistry* **41**, 653–658, 2006.
62. C. D. Skory, S. N. Freer, and R. J. Bothast. Screening for ethanol-producing filamentous fungi, *Biotechnology Letters* **19**, 203–206, 1997.
63. T. L. Lubbehusen, J. Nielsen, and M. McIntyre. Aerobic and anaerobic ethanol production by *Mucor circinelloides* during submerged growth, *Applied Microbiology and Biotechnology*, **63**, 543–548, 2004.
64. A. Sues, R. Millati, L. Edebo, and M. J. Taherzadeh. Ethanol production from hexoses, pentoses, and dilute-acid hydrolyzate by *Mucor indicus*, *FEMS Yeast Research* **5**, 669–676, 2005.
65. K. Karimi, T. Brandberg, L. Edebo, and M. Taherzadeh. Fed-batch cultivation of *mucor indicus* in dilute-acid lignocellulosic hydrolyzate for ethanol production, *Biotechnology Letters* **6**, 1395–1400, 2005.
66. P. A. Gibbs, R. J. Seviour, and F. Schmid. Growth of filamentous fungi in submerged culture: Problems and possible solutions, *Critical Reviews in Biotechnology* **20**, 17–48, 2000.
67. M. Papagianni. Fungal morphology and metabolite production in submerged mycelial processes, *Biotechnology Advances* **22**, 189–259, 2004.
68. J. N. de Vasconcelos, C. E. Lopes, and F. P. de França. Continuous ethanol production using yeast immobilized on sugar-cane stalks, *Brazilian Journal of Chemical Engineering* **21**, 357–365, 2004.
69. V. M. da Matta and R. A. Medronho. A new method for yeast recovery in batch ethanol fermentations: Filter aid filtration followed by separation of yeast from filter aid using hydrocyclones, *Bioseparation* **9**, 43–53, 2000.
70. B. McNiel and L. M. Harvey. Fermentation a practical approach, In: *Practical Approach Series*, Hames, B. D. (Ed.), First ed., Oxford: Oxford University Press, 1990.
71. Y. Harada, K. Sakata, S. Sato, and S. Takayama. Fermentation Pilot Plant, 1997.
72. G. R. Cysewski and C. R. Wilke. Process design and economic studies of alternative fermentation methods for the production of ethanol, *Biotechnology and Bioengineering* **20**, 1421–1444, 1978.
73. L. Garlick. Biostil—Fermentation of molasses cane juice using continuous fermentation, *Sugar Journal* **46**(4), 13–16, 1983.
74. F. Talebnia and M. J. Taherzadeh. In situ detoxification and continuous cultivation of dilute-acid hydrolyzate to ethanol by encapsulated *Saccharomyces cerevisiae*, *Journal of Biotechnology* **125**, 377–384, 2006.
75. T. Ghose and R. Tyagi. Rapid ethanol fermentation of cellulose hydrolyzate. II. Product and substrate-inhibition and optimization of fermentor design, *Biotechnology and Bioengineering* **21**, 1401–1420, 1979.
76. K. Rosen. Continuous production of alcohol, *Process Biochemistry* **13**, 25–26, 1978.
77. R. Purwadi. Continuous ethanol production from dilute-acid hydrolyzates: Detoxification and fermentation strategy, Department of Chemical and Biological Engineering, *Chemical Reaction Engineering,* Chalmers University of Technology, Göteborg, Sweden, 2006.
78. C. S. Chen. *Hawaii Ethanol from Molasses Project—Final Report*, HNEI-80-03, Honolulu, Hawaii: Hawaii Natural Energy Institute, University of Hawaii at Manoa, 1980.
79. I. Karaki, M. Konishi, K. Amakai, and Ishikava. Alcohol production by continuous fermentation of molasses, *Hakko Kyokaishi* **30**, 106, 1972.

80. G. P. Philippidis, T. K. Smith, and C. E. Wyman. Study of the enzymatic hydrolysis of cellulose for production of fuel ethanol by the simultaneous saccharification and fermentation process, *Biotechnology and Bioengineering* **41**, 846–853, 1993.
81. T. Uragami, K, Okazaki, H. Matsugi, and T. Miyata. Structure and permeation characteristics of an aqueous ethanol solution of organic-inorganic hybrid membranes composed of poly(vinyl alcohol) and tetraethoxysilane, *Macromolecules* **35**, 9156–9163, 2002.
82. A. C. Wilkie, K. J. Riedesel, and J. M. Owens. Stillage characterization and anaerobic treatment of ethanol stillage from conventional and cellulosic feedstocks, *Biomass and Bioenergy* **19**, 63–102, 2000.
83. V. Dubey, L. K. Pandey, and C. Saxena. Pervaporative separation of ethanol/water azeotrope using a novel chitosan-impregnated bacterial cellulose membrane and chitosan–poly(vinyl alcohol) blends, *Journal of Membrane Science* **251**, 131–136, 2005.
84. Z. Changluo, L. Moe, and X. We. Separation of ethanol-water mixtures by pervaporation-membrane separation process, *Desalination* **62**, 299–313, 1987.

Raw Materials to Produce Low-Cost Biodiesel

M. P. Dorado

4.1 Introduction

The present energy scenario is undergoing a period of transition, as more and more energy consumers understand the inevitability of exhaustion of fossil fuel. The era of fossil fuel of nonrenewable resources is gradually coming to an end, where oil and natural gas will be depleted first, followed eventually by depletion of coal. In developing countries, the energy problem is rather critical. The price paid for petrol, diesel, and petroleum products now dominates over all other expenditures and forms a major part of a country's import bill. In view of the problems stated, there is a need for developing alternative energy sources. Alternative fuel options are mainly biogas, producer gas, methanol, ethanol, and vegetable oils. But biogas and producer gas have low energy contents per unit mass and can substitute for diesel fuel only up to 80%. Moreover, there are problems of storage because of their gaseous nature. Methanol and ethanol have very poor calorific value per unit mass, apart from having a low Cetane number. Therefore, these are rather unsuitable as substitutes for high-compression diesel engines. Experimental evidence indicates that methanol and ethanol can be substituted up to only 20–40%. There exists a number of plant species that produce oils and hydrocarbon substances as a part of their metabolism. These products can be used with other fuels in diesel engines with various degrees of processing. The development of vegetable oils as liquid fuels have several advantages over other alternative fuel options, such as

1. The technologies for extraction and processing are very easy and simple, as conventional equipment with low energy inputs are needed.

2. Fuel properties are close to diesel fuel.

3. Vegetable oils are renewable in nature.

4. Being liquid, these oils offer ease of portability and also possess stability and no handling hazards.

5. The by-product leftover after extraction of oil is rich in protein and can be used as animal feed or solid fuel.

6. Cultivation of these oilseeds is adaptable to a wide range of geographic locations and climatic conditions.

7. Biodiesel can be used directly in compression ignition engines with no substantial modifications to the engine.

8. Biodiesel contains no sulfur, and there is no production of oxides of sulfur.

Hence, biodiesel is considered an alternative fuel for internal combustion engines derived from oils and fats from renewable biological sources; it emits far less regulated pollutants than the standard diesel fuel [1–10]. It entails minimal reduced engine performance as a result of a slight power loss and increase in specific fuel consumption [8, 11–29]. However, one main concern in further usage of biodiesel is the economic viability of producing biodiesel.

The main economic criteria are manufacturing cost and the price of raw feedstock. Manufacturing costs include direct costs for oil extraction, reagents, and operating supplies, as well as indirect costs related to insurance and storage. Fixed capital costs involved in the construction of processing plants and auxiliary facilities, distribution, and retailing must also be taken into consideration [30].

Several authors have found that biodiesel is currently not economically feasible unless tax credits are applied [23, 31]. Peterson [23] has found that diesel fuel costs less than biodiesel, and an emergency or diesel shortage would be required to provide a practical reason for using biodiesel. Some authors have stated that biodiesel could compete with diesel fuel if produced in cooperatives [31, 32].

To promote biodiesel consumption, several countries have exempted biodiesel from their fuel excise tax. Among them, the European Union (EU) approved the biodiesel tax exemption program in May 2002. The financial law funded biofuels through excise exemption over a period of 3 years (Art. 21, Finance Law 2001). The U.S. Senate Finance Committee also approved an excise tax exemption for biodiesel in 2003. Moreover, the legislation provides a 1% reduction in the diesel fuel excise tax for each percentage of biodiesel blended with petroleum diesel up to 20%. Also, among some other countries, the Australian Senate approved an excise exemption on biofuels in 2004. However, the tax exemption will one day come to an end; in order to continue to promote the social inclusion and economic attraction of biodiesel, other steps will be needed. This could be facilitated by the selection of low-cost raw materials, such as nonedible oils, used frying oil, or animal fat, and the use of a lower-cost transesterification process.

A lower-cost biodiesel production can be achieved by the optimization of the process. Because the chemical properties of the esters determine their feasibility as a fuel, the intent of the optimization is to investigate and optimize the involved parameters maximizing the yield of ester, to develop a low-cost chemical process, and to ensure appropriate oil chemical properties for both the transesterification and the engine performance.

Although it is a well-known process since, in 1864, Rochleder described glycerol preparation through the ethanolysis of castor oil [33], the proportion of reagents affects the process, in terms of conversion efficiency [34]; this factor differs according to the vegetable oil. Several researchers have identified the most important variables that influence the transesterification reaction, namely, reaction temperature, type and amount of catalyst, ratio of alcohol to vegetable oil, stirring rate, and reaction time [20, 35–42]. In this sense, it is important to characterize the oil (i.e., fatty acid composition, water content, and peroxide value) to determine the correlation between them and the feasibility to convert the oil into biodiesel [39, 43].

However, several studies have identified that the price of feedstock oils is by far one of the most significant factors affecting the economic viability of biodiesel manufacture [30, 44–46]. Approximately 70–95% of the total biodiesel production cost arises from the cost of the raw material [44, 45]. To produce a competitive biodiesel, the feedstock price is a factor that needs to be taken into consideration. Edible oils are too valuable for human feeding to run automobiles. So, the accent must be on nonedible oils and used frying oils.

4.2 Nonedible Oils

Among nonedible feedstock, there are many crops and tree-borne oilseed plants, such as karanja, neem, and jatropha, which have been underutilized due to the presence of toxic components in their oils. Most of them grow in underdeveloped and developing countries, where a biodiesel program would give multiple benefits in terms of generation of employment for rural people (farmers), leverage of starting many types of industries using by-products from biofuels, and so forth [47]. However, nonedible crops are very much ignored in most cases. They grow on, regardless, waiting for their energetic potential to be discovered. The key is to find crops or trees that need very little care, have high oil content, and are resistant to plagues and drainage. The foliage could be used as manure, giving an added value to the crop. In fact, most of the trees and crops mentioned in the following (karanja, neem, etc.) grow well on wasteland and can tolerate long periods of drought and dry conditions.

4.2.1 Bahapilu oil

Crop description. *Salvadora oleoides* Decne, *S. persica* L., and *S. indica*—commonly known as bahapilu, chootapilu, jhal, jaal, pilu, kabbar, khakan, and mitijar—belong to the family Salvadoraceae and are found in arid regions of western India and Pakistan (see Figs. 4.1 and 4.2). The crop is typical of the tropical thorn forest. It is highly salt tolerant and grows in coastal regions and on inland saline soils [48, 49]. *S. oleoides* is a shrub or small tree up to 9 m in height. Seeds contain 40–50% of a greenish-yellow fat containing large amounts of lauric and myristic acids [50].

Figure 4.1 *Salvadora persica.* (*Photo courtesy of Abdulrahman Alsirhan [www.alsirhan.com/Plants_s/Salvadora_persica.htm].*)

Figure 4.2   *Salvadora angustifolia. (Photo courtesy of Dr. Kazuo Yamasaki [http://pharm1.pharmazie.uni-greifswald.de/gallery/gal-salv.htm].)*

Main uses.   The fruits are sweet and edible. The seed cake contains 12% protein and is suitable for livestock fodder. The wood is used for building purposes. It is also an important source of fuelwood. The fat in seeds can be used for making soap and candles. The leaves and fruits are used in medicine to relieve cough, rheumatism, and fever. The tree contributes to erosion control in fragile areas [50]. Some authors have carried out systematic studies on the lubrication properties of biodiesel from *S. oleoides* and its blends. Biodiesel was prepared by base-catalyzed transesterification using methanol. Results indicate that addition of biodiesel improves the lubricity and reduces wear scar diameter even at a 5% blend [51].

4.2.2   Castor oil

Crop description.   *Ricinus communis* L., commonly known as the castor-oil plant, belongs to the family Euphorbiaceae (see Fig. 4.3). This perennial tree or shrub can reach up to 12 m high in tropical or subtropical climates, but it remains 3 m tall in temperate places. Native to Central Africa, it is being cultivated in many hot climates. The oil contains up to 90% ricinoleic acid, which is not suitable for nutritional purposes due to its laxative effect [52]. This hydroxycarboxylic acid is responsible for the extremely high viscosity of castor oil, amounting to almost a hundred times the value observed for other fatty materials [53].

Figure 4.3 *Ricinus communis* L. (*Photo courtesy of Eric Winder [www.bio.mtu.edu/~jclewin/bahama_pics/Eric/].*)

Main uses. Castor bean is cultivated for its seeds, which yield a fast-drying oil used mainly in industry and medicine. Coating fabrics, high-grade lubricants, printing inks, and production of a polyamide nylon-type fiber are among its uses. Dehydrated oil is an excellent drying agent and is used in paints and varnishes. Hydrogenated oil is utilized in the manufacture of waxes, polishes, carbon paper, candles, and crayons. The pomace or residue after crushing is used as a nitrogen-rich fertilizer. Although it is highly toxic due to the ricin, a method of detoxicating the meal has been developed, so that it can safely be fed to livestock [54]. Several authors have found that castor-oil biodiesel can be considered as a promising alternative to diesel fuel. Transesterification reactions have been carried out mainly by using both ethanol and NaOH, and through enzymatic methanolysis [55–57]. Several authors have studied the influence of the nature of the catalyst on the yields of biodiesel from castor oil. They found that the most efficient transesterification of castor oil could be achieved in the presence of methoxide and acid catalysts [58]. The influence of alcohol has also been studied. Comparing the use of ethanol versus methanol, Meneghetti et al. have found that similar yields of fatty acid esters may be obtained; however, the reaction with methanolysis is much more rapid [59]. Cvengros et al. produced both ethyl and methyl esters, using NaOH in the presence of ethanol and methanol, respectively. Despite the high viscosity and density values, they concluded that both methyl and ethyl esters can be successfully used as fuel. A positive solution to meet the standard values for both viscosity and density parameters can be a dilution with esters based on oils/fats without an OH group, or a blending with conventional diesel fuel [60].

4.2.3 Cottonseed oil

Crop description. *Gossypium* spp., commonly known as cotton, belongs to the family Malvaceae and is native to the tropical and subtropical regions (see Fig. 4.4). Four separate domesticated species of cotton grown in various parts of the world are *G. arboreum* L., *G. herbaceum* L., *G. hirsutum* L., and *G. barbadense* L. Cotton shrubs are annual and found in the United States, Australia, Asia, and Egypt. Some have been grown for many years in southern Europe, mainly the Balkans and Spain. It can grow up to 3 m high [61–64].

Main uses. Cotton is a major world fiber crop. Its fiber grows around the seeds of the cotton plant and is used to make textile, which is the most widely used natural-fiber cloth. The seeds yield a valuable oil used for the production of cooking oil or margarine. The fatty acid composition includes mainly palmitic acid (21%), stearic acid (2.4%), oleic acid (19.5%), linoleic acid (54.3%), and myristic acid (0.9%). Cottonseed oil, cake, meal, and hulls for feeding are other uses of the by-products. Whole cottonseed may be used as a feed for mature cattle. Cottonseed meal is an excellent protein supplement for cattle. The limitations on effective utilization of this product in rations for swine and poultry are of minor significance to ruminant animals. Cottonseed meal has a relatively low rumen degradability and is therefore a good source of by-pass protein and is especially useful in rations for milking cows [61–64].

Figure 4.4 *Gossypium* spp. (*Photo courtesy of Prof. Jack Bacheler [http://ipm.ncsu.edu/ cotton/InsectCorner/photos/ images/Open_cotton_plant.jpg].*)

Köse et al. investigated the transesterification of refined cottonseed oil, using primary and secondary alcohols (oil–alcohol molar ratio 1:4) in the presence of an immobilized enzyme from *Candida antarctica* (30% enzyme, based on oil weight). The reaction was carried out at 50°C for 7 h, showing that conversion using secondary alcohols was more effective [65]. Some authors have also proposed the use of lipase with methanol [66]. Royon et al. used the same catalyst in a *t*-butanol solvent. Maximum yield was observed after 24 h at 50°C with a reaction mixture containing 32.5% *t*-butanol, 13.5% methanol, 54% oil, and 0.017 g of enzyme per g of oil [67]. Recent tendencies propose the use of ultrasonically assisted extraction transesterification to increase ester yield [68].

4.2.4 Cuphea oil

Crop description. *Cuphea* spp., *C. carthagenensis*, *C. painter*, *C. ignea*, and *C. viscosissima*—commonly known as cuphea—belong to the family Lythraceae and grow in temperate and subtropical climates (see Fig. 4.5). They can be found in Central and South America, and have been grown in trials in Germany and the United States. The seeds of *Cuphea* contain about 30–36% oil [69]. Major fatty acid composition of the oil includes caprylic acid (73% in *C. painter*, 3% in *C. ignea*), capric acid (18% in *C. carthagenensis*, 24% in *C. painteri*, 87% in *C. ignea*, and 83–86% in *C. llavea*), and lauric acid (57% in *C. carthagenensis*) [70].

Main uses. It contains high levels of short-chain fatty acids, very interesting for industrial applications. Previous studies have suggested that oil composition and chemical properties of *C. viscosissima* VS-320 are not appropriate for use as a substitute for diesel fuel without chemical

Figure 4.5 *Cuphea* sp. (*Photo courtesy of Dr. Alvin R. Diamond [http://spectrum.troy.edu/~diamond/PIKEFLORA.htm].*)

conversion of triglycerides to methyl esters. Further genetic modifications must be made [71, 72]. Later studies have revealed that genetically modified oils present relatively low viscosity that is predicted to enhance their performance as alternative diesel fuels [73]. Also, atomization properties suggest better fuel performance, because this oil has short-chain triglycerides, while traditional vegetable oils contain predominantly long-chain triglycerides [74].

4.2.5 *Jatropha curcas* oil

Crop description. *J. curcas*—commonly known as pourghere, ratanjyot, Barbados nut, physic nut, parvaranda, taua taua, tartago, saboo dam, jarak butte, or awla—belongs to the family Euphorbiaceae and grows in hot, dry, tropical climates (see Fig. 4.6). It originated from South America and is now found worldwide in tropical countries. It grows wild especially in West Africa, and is grown commercially in the Cape Verde Islands and Malagasy Republic. The tree reaches a height of 8 m and is a tough, drought-resistant plant that bears oil-rich seeds prolifically under optimum growing conditions [75]. The seeds contain about 55% oil [76]. The oil contains a toxic substance, curcasin, which has a strong purging effect. Major fatty acid composition consists of myristic acid (0–0.5%), palmitic acid (12–17%), stearic acid (5–6%), oleic acid (37–63%), and linoleic acid (19–40%) [77].

Main uses. It has been cultivated as a drought-resistant plant in marginal areas to prevent soil erosion. The oil has been commercially used

Figure 4.6 *Jatropha curcas. (Photo courtesy of Piet Van Wyk and EcoPort [www.ecoport.org].)*

for lighting purposes, as lacquer, in soap manufacture, and as a textile lubricant. It is also used for medicinal purposes for its strong purging effect. The leaves are used in the treatment of malaria. Products useful as plasticizer, hide softeners, and hydraulic fluid have been obtained after halogenation [75]. The wood is used for fuel. The cake, after oil extraction, cannot be used for animal feed due to its toxicity, but is a good organic fertilizer. The wood is very flexible and is used for basket making. A water extract of the whole plant has molluscicide effects against various types of snail, as well as insecticidal properties [77].

Recently, there has been considerable interest in the use of the oil in small diesel engines. This oil has great potential for biodiesel production [78–80]. Foidl et al. transesterified *J. curcas* oil, using a solution of KOH (0.53 mol) in methanol (10.34 mol) and stirring at 30°C for 30 min [81]. The ester fuel has high quality and meets the existing standards for vegetable-oil-derived fuels. Some researchers have proposed the use of immobilized enzymes such as *Chromobacterium viscosum*, *Candida rugosa*, and *Porcine pancreas* as a catalyst [82, 83]. Modi et al. have proposed the use of propan-2-ol as an acyl acceptor for immobilized *Candida antarctica* lipase B. Best results have been obtained by means of 10% Novozym-435 based on oil weight, with a alcohol–oil molar ratio of 4:1 at 50°C for 8 h [84]. Zhu et al. have proposed the use of a heterogeneous solid superbase catalyst (catalyst dosage of 1.5%) and calcium oxide, at 70°C for 2.5 h, with a methanol–oil molar ratio of 9:1 to produce biodiesel [85]. The lubrication properties of this biodiesel have also been taken into consideration [51].

4.2.6   Karanja seed oil

Crop description.   *Pongamia pinnata* (L.) Pierre, *P. glabra* Vent., *Cytisus pinnatus* L., *Derris indica* (Lam.) Bennett, and *Galedupa indica* Lam.— commonly known as karanja, pongam, coqueluche, Vesi Ne Wai, vesivesi, hongay, and honge—belong to the Leguminaceae family and are widely distributed in tropical Asia (see Fig. 4.7). The tree is drought-resistant, tolerant to salinity, and is commonly found in East Indies, Philippines, and India. The karanja tree grows to a height of about 1 m and bears pods that contain one or two kernels. The kernel oil content varies from 27% to 39% and contains toxic flavonoids, including 1.25% karanjin and 0.85% pongamol [86–88]. The fatty acid composition consists of oleic acid (44.5–71.3%), linoleic acid (10.8–18.3%), palmitic acid (3.7–7.9%), stearic acid (2.4–8.9%), and lignoceric acid (1.1–3.5%) [86, 89].

Main uses.   The oil is used mainly in agriculture, pharmacy (particularly in the treatment of skin diseases), and the manufacture of soaps. It has insecticidal, antiseptic, antiparasitic, and cleansing properties, like neem oil [86–88]. The cake after oil extraction may be used as manure.

Figure 4.7 *Pongamia pinnata* (L.) Pierre. (*Photo courtesy of the Food and Agricultural Organization of the United Nations [www.fao.org].*)

All parts of the plant have also been analyzed for its reported medical importance. Several scientists have investigated and guaranteed karanja oil as a potential source of biodiesel [78]. Most researchers have conducted the transesterification of *P. pinnata* oil by using methanol and potassium hydroxide catalysts [90–92]. Meher et al. [90] found that using a methanol–oil molar ratio of 12:1 produced maximum yield of biodiesel (97%), while Vivek and Gupta [91] stated the optimum ratio was 8–10:1. In both cases, the optimal temperature was around 65°C, with a reaction time of 180 min [90] and 30–40 min [91]. Vivek and Gupta used 1.5% w/w of catalyst (KOH), while Meher et al. used 2% w/w solid basic Li/CaO catalyst [93]. Due to the high FFA (free fatty acid) content, some researchers have proposed esterification with H_2SO_4 prior to transesterification with NaOH [94, 95]. In all cases, karanja oil has shown a feasibility to be used as a raw material to produce biodiesel, saving large quantities of edible vegetable oils. Diesel engine performance tests were carried out with karanja methyl ester (KME) and its blend with diesel fuel from 20% to 80% by volume [92]. Results have revealed a reduction in exhaust emissions together with an increase in torque, brake power, thermal efficiency, and reduction in brake-specific fuel consumption, while using the blends of karanja-esterified oil (20–40%), compared to straight diesel fuel.

4.2.7   Linseed oil

Crop description.   *Linum usitatissimum* L.—commonly known as linseed, flaxseed, lint bells, or winterlien—belongs to the family Linaceae (see Fig. 4.8). This annual herb can grow up to 60 cm in height in most temperate and tropical regions. This plant is native to West Asia and the Mediterranean [96]. The seeds contain 30–40% oil, including palmitic

Figure 4.8   *Linum usitatissimum.*

acid (4.5%), stearic acid (4.4%), oleic acid (17.0%), linoleic acid (15.5%), and linolenic acid (58.6%).

Main uses.   Medicinal properties of the seeds have been known since ancient Greece. It is used in pharmacology (antitussive, gentle bulk laxative, relaxing expectorant, antiseptic, antiinflammatory, etc.) [97]. As the source of linen fiber, it was used by the Egyptians to make cloth in which to wrap their mummies. However, today it is mainly grown for its oil [98, 99], which is used in the manufacture of paints, varnishes, and linoleum. Linseed oil is used as a purgative for sheep and horses. It is also used in cooking. There is a market for flaxseed meal as both animal feeding and human nutrition [96].

Lang et al. transesterified linseed oil by using different alcohols (methanol, ethanol, 2-propanol, and butanol) and catalysts (KOH and sodium alkoxides). Butyl esters showed reduced cloud points and pour points [100]. Some authors have found that biodiesel from linseed oil presents a lower cold filter plugging point (CFPP) than biodiesel from rapeseed oil, due to large amounts of linolenic acid methyl ester and their iodine value [101]. Long-term endurance tests have been carried out with methyl esters of linseed oil, showing low emission characteristics. Wear assessment has shown lower wear for a biodiesel-operated engine [102]. Experimental investigations on the effect of 20% biodiesel blended with diesel fuel on lubricating oil have shown a lubricating oil life longer while operating the engine on biodiesel [103]. Oxidation stability have shown better results compared with methyl esters of animal origin [104]. Lebedevas et al. have suggested the use of three-component mixtures (rapeseed-oil methyl esters, animal methyl esters, and linseed oil methyl esters) to fuel the engine. These three-component mixtures reduced exhaust emissions significantly, with the exception of NO_x that increased them up to 13% [105].

4.2.8 Mahua oil

Crop description. *Madhuca indica*—commonly known as madhuka, yappa, mahuda, mahua, mauwa, mohwa, hippe, butter tree, mahwa, mahula, or elupa—belongs to the family Sapotaceae and grows up to 21 m high. This deciduous tree is distributed mainly in India (see Fig. 4.9). The kernels are 70% of seed by weight. Seeds content includes 35% oil and 16% protein. Main fatty acids are palmitic acid (16–28.2%), stearic acid (20–25.1%), arachidic acid (3.3%), oleic acid (41–51%), and linoleic acid (8.9–13.7%) [106].

Main uses. Traditionally, it has been used as a source of natural hard fat in soap manufacture. The seed oil is used as an ointment in rheumatism and to prevent dry, cracked skin in winter. It is used in foods, cosmetics, and lighting. The cake presents toxic and bitter saponins that preclude its use as animal feed. However, mahua cake can be used as organic manure [106]. Several approaches to produce biodiesel can be found. Ghadge and Raheman have proposed a two-step pretreatment to reduce high FFA levels. Transesterification was carried out adding 0.25 v/v methanol and 0.7% KOH. Fuel properties were found comparable to those of diesel fuel [107]. Some authors have proposed different successful

Figure 4.9 *Madhuca indica. (Photo courtesy of Antonie van den Bos [www.botanypictures.com/plantimages/].)*

alternatives to produce biodiesel: ethanol and sulfuric acid, and methanol and NaOH [108–110]. Puhan et al. have found better diesel engine performance for methyl esters compared to ethyl and butyl esters, while ethyl esters show lower NO_x emissions compared to the rest [111]. Systematic studies on the lubrication properties of biodiesel have shown that the preferred range of blending with diesel fuel is 5–20% [51].

4.2.9 Nagchampa oil

Crop description. *Calophyllum inophyllum*—commonly known as nagchampa, ballnut, ati tree, kamani, ndamanu, fetau, Alexandrian laurel, nambagura, Indian laurel, and tamanu oil—belongs to the family Guttiferae and is native to the Indo–Pacific region, particularly Malaysia [112]. This evergreen tree is commonly found in the coastal regions of South India and Madagascar (see Fig. 4.10). It usually reaches up to 25 m high [113]. It tolerates varied kinds of soil, coastal sand, clay, and degraded soil. The average kernel oil content is about 60.1% [114]. The fatty acids present in crude oils are stearic (14.3%), palmitic (13.7%), oleic (39.1%), and linoleic (31.1%) acids [115].

Main uses. It is known best as an ornamental plant. Besides this, its wood is hard and strong and has been used in construction. The seeds yield oil for medicinal use and cosmetics. A number of medicinal and therapeutic properties of various parts of *Calophyllum* have been described, including the treatment of rheumatism, varicose veins, hemorrhoids, and chronic ulcers [116]. Fatty acid methyl esters of *C. inophyllum* oil have been found suitable for use as biodiesel that meets biodiesel standards of the United States and European Standards Organization [78].

Figure 4.10 *Calophyllum inophyllum.* (*Photo by Forest Starr and Kim Starr, courtesy of the U.S. Geological Society [www.hear.org/ starr/hiplants/images/hires/ html/starr_040711_0232_ calophyllum_inophyllum.htm].*)

4.2.10   Neem oil

Crop description.   *Azadirachta indica*—commonly known as the neem tree, nim, margosa, veppam, cho do, or nilayati nimb—belongs to the family Meliaceae and can be found in dry tropical forests (see Fig. 4.11). The major producing countries are India, Sri Lanka, Burma, Pakistan, tropical Australia, and Africa. The evergreen neem tree grows up to 18 m high. The fat content of the kernels ranges from 33 to 45% [77]. The fatty acid content includes 42% oleic acid, 20% palmitic acid, 20% stearic acid, 15% linoleic acid, and 1.4% arachidic acid. Good quality kernels yield 40–50% oil. The cakes, which contain 7–12% oil are sold for solvent extraction. Neem oil is unusual in that it contains nonlipid associates often loosely termed as bitters and organic sulfur compounds that impart a pungent, disagreeable odor [88].

Main uses.   The products of the neem tree are known to be antibacterial, antifungal, and antiparasitic. The main uses are in soaps, teas, medicinal preparations, cosmetics, skin care, insecticides, and repellents. Neem twigs are used as tooth brushes and ward against gum disease. Neem oil, which is extracted from the seed kernel, has excellent healing properties and is used in creams, lotions, and soaps. It is also an effective fungicide. The bitter cake after the extraction of oil

Figure 4.11   *Azadirachta indica. (Photo courtesy of Food and Agricultural Organization of the United Nations [www.fao.org].)*

has no value for animal feeding but is recognized as both a fertilizer and nematicide [88]. Besides medical use, esters of neem oils have some important fuel properties that can be exploited for alternative fuels for diesel engines [78]. Nabi et al. have produced biodiesel from neem oil by using 20% methyl alcohol and 0.6% anhydrous lye catalyst (NaOH). The temperature of the materials was maintained at 55–60°C. Compared with conventional diesel fuel, exhaust emissions including smoke and CO were reduced, while NO_x emission was increased with diesel–biodiesel blends. However, NO_x emission with diesel–biodiesel blends was slightly reduced when exhaust gas recirculation (EGR) was applied. According to the results, Nabi et al. have recommended the use of the ester of this oil as an environment-friendly alternative fuel for diesel engines [117].

4.2.11   Rubber seed oil

Crop description.   *Hevea brasiliensis*—commonly known as Pará rubber tree, rubber tree, jebe, cauchotero de para, seringueira, or siringa—belongs to the family Euphorbiaceae (see Fig. 4.12). The rubber tree originates from the Amazon rain forest (Brazil). Today, most rubber tree plantations are located in Southeast Asia and some are also in tropical Africa. The tree can reach up to 30 m high. Oil can be extracted from the seeds. Although there are variations in the oil content of the seed from different countries, the average oil yield has been 40% [118].

Figure 4.12   *Hevea brasiliensis.*

Its fatty acid composition includes palmitic acid (10.2%), stearic acid (8.7%), oleic acid (24.6%), linoleic acid (39.6%), and linolenic acid (16.3%) [118].

Main uses. The crop is of major economical importance because it produces latex. The wood from this tree is used in the manufacture of high-end furniture. In Cambodia and other rubber-manufacturing countries, rubber seeds are used to feed livestock. Rubber seed contains cyanogenic glycosides that will release prussic acid in the presence of enzymes or in slightly acidic conditions. Press cake or extracted meal can be cautiously used as fertilizer or feed for stock [119].

Several studies to check its feasibility as a source of biodiesel have been undertaken. Ikwuagwu et al. have prepared methyl esters of rubber seed oil using excess of methanol (6 M) containing 1% NaOH as a catalyst. Petroleum ether was added to the reaction to produce two phases. Analysis of the properties have shown a good potential for use as an alternative diesel fuel, with the exception of the oxidative stability [120]. Ramadhas et al. have performed a previously acid-catalyzed esterification to reduce the high FFA content, followed by an alkaline esterification. Sulfuric acid 0.5% by volume and a methanol-oil molar ratio of 6:1 was used in the pretreatment. A molar ratio of 9:1 and 0.5% by weight of sodium hydroxide was used during the second step. The authors found a reduction in exhaust gas emissions. The lower blends of biodiesel increased brake thermal efficiency and reduced fuel consumption [121].

4.2.12   Tonka bean oil

Crop description. *Dipteryx odorata*—commonly known as sarapia, tonka bean, amburana, aumana, yape, charapilla, and cumaru—belongs to the family Leguminacea and grows in tropical areas (see Fig. 4.13). Major producing countries are Guianas and Venezuela. The tonka bean is the seed of a large tree. The kernel contains up to 46% oil on a dry basis. Major fatty acid composition of oil includes palmitic acid (6.1%), stearic acid (5.7%), oleic acid (59.6%), and linoleic acid (51.4%) [77].

Main uses. The oil is used in perfumery and as a flavoring material. Tonka extracts are used in the tobacco industry to impart a particular aroma. Few attempts have been made to use it as a raw material to produce biodiesel. Abreu et al. conducted methanolysis of cumaru oil using different homogeneous metal (Sn, Pb, and Zn) complexes as catalysts. They found that pyrone complexes of different metals are active for cumaru-oil transesterification reaction [122].

Figure 4.13 *Dipteryx odorata. (Photo courtesy of Dr. Davison Shillingford [www.da-academy.org/dagardens_tonkabean1.html].)*

4.3 Low-Cost Edible Oils

Besides nonedible oils, there are some edible oils from plants that yield a relatively lower-cost source to produce biodiesel compared to biodiesel from rapeseed oil or soybean oil.

4.3.1 Cardoon oil

Crop description.   *Cynara cardunculus* L.—commonly known as cardoon, Spanish artichoke, artichoke thistle, cardone, dardoni, or cardo—belongs to the family Asteracea (see Fig. 4.14). Artichokes originated in the Mediterranean region and climates, becoming an important weed of the Pampas in Argentina, and in Australia, and California because of its adaptation to dry climate. Its fatty acid composition mainly includes palmitic acid (19.3%), stearic acid (6.1%), oleic acid (39%), and linoleic acid (30%) [123].

Main uses.   The leaf stalks are eaten as a vegetable. The leaves contain cynarin, which improves gall bladder and liver functions, increases bile flow, and lowers cholesterol. The down from the seed heads is used as rennet.

Encinar et al. transesterified *C. cardunculus* oil using methanol and several catalysts (sodium hydroxide, potassium hydroxide, and sodium methoxide) to produce biodiesel. Best properties were achieved by using 15% methanol and 1% sodium methoxide as catalyst, at 60°C temperature [124].

Figure 4.14   *Cynara cardunculus.*

The reaction can also be accomplished by using an ethanol–oil molar ratio of 12:1 and 1% sodium hydroxide, at 75°C [125]. *C. cardunculus* methyl esters also provide a significant reduction in particulate emissions, mainly due to reduced soot and sulfate formation [126].

4.3.2   Ethiopian mustard oil

Crop description.   *Brassica carinata*, commonly known as Ethiopian mustard, is an adequate oil-bearing crop that is well adapted to marginal regions (see Fig. 4.15). This crop, which is originally from Ethiopia, is drought-resistant and grown in arid regions [127, 128]. Ethiopian mustard presents up to 6% saturated hydrocarbon chains. It is native to the Ethiopian highlands, is widely used as food by the Ethiopians, and pre-sents better agronomic performances in areas such as Spain, California, and Italy. This makes *B. carinata* a promising oil feedstock for cultivation in coastal Mediterranean areas, which could offer the possibility of exploiting the Mediterranean marginal areas for energetic purposes [129]. Its fatty acid composition includes palmitic acid (3.6%), stearic acid (1.3%), oleic acid (14.8%), linoleic acid (12.2%), gadoleic acid (10.3%), and erucic acid (45.4%) [123].

Main uses.   It is widely used as food in Ethiopia. Oil from wild species is high in erucic acid, which is toxic, although some cultivars contain little erucic acid and can be used as food. The seed can also be crushed and used as a condiment [127]. There is a genetic relationship among

Figure 4.15 *Brassica carinata.*

B. carinata genotypes based on oil content and fatty acid composition. Genet et al. have generated information to plan crosses and maximize the use of genetic diversity and expression of heterosis [130]. Dorado et al. found negative effects of singular fatty acids, such as erucic acid, over alkali-catalyzed transesterification reaction [39]. These researchers described a low-cost transesterification process of *B. carinata* oil. An oil–methanol molar ratio of 1:4.6, addition of 1.4% of KOH, a reaction temperature in the range of 20–45°C, and 30 min of stirring are considered to be the best conditions to develop a low-cost method to produce biodiesel from *B. carinata* oil [39, 131]. Biodiesel from Ethiopian mustard oil could become of interest if a fuel tax exemption is granted [30]. When compared with petroleum diesel fuel, Cardone et al. have found that engine test bench analysis did not show any appreciable variation of output engine torque values, while there was a significant difference in specific fuel consumption data at the lowest loads. Biodiesel produced higher levels of NO_x concentrations and lower levels of particulate matter (PM), with respect to diesel fuel. Biodiesel emissions contain less soot [132].

4.3.3 Gold-of-pleasure oil

Crop description.   *Camelina sativa* L. Crantz—commonly known as gold-of-pleasure and camelina—belongs to the family Cruciferae and grows well in temperate climates (see Fig. 4.16). It is an annual oilseed plant and is cultivated in small amounts in France, and to a lesser

Figure 4.16 *Camelina sativa* L. Crantz. *(Photo courtesy of Prof. Arne Anderberg [http://linnaeus. nrm.se/flora/di/brassica/camel/ camemic.html].)*

extent in Holland, Belgium, and Russia. The oil content of camelina seeds ranges from 29.9% to 38.3%. However, it is an underexploited oilseed crop at present. Its fatty acid profile includes oleic acid (14–19.5%), linoleic acid (18.8–24%), linolenic acid (27–34.7%), eicosenoic acid (12–15%), and erucic acid (less than 4%) [133]. Budin et al. have concluded that camelina is a low-input crop possessing a potential for food and nonfood exploitation [133].

Main uses.   This crop has recently been rediscovered as an oil crop. At the moment, the feasibility of utilizing oil from this plant is being investigated [53, 134]. Oil is used as a luminant and emollient for softening the skin. Fiber is obtained from the stems. Fröhlich and Rice have investigated production of methyl ester from camelina oil. Biodiesel was prepared by means of a single-stage esterification using methanol and KOH [135]. Steinke et al. have developed both alkali-catalyzed and lipase-catalyzed alcoholyses of camelina oil [136, 137].

4.3.4   Tigernut oil

Crop description.   *Cyperus esculentus* L.—commonly known as tigernut, chufa sedge, yellow nutsedge, and earth almond—belongs to the family Cyperaceae and grows in warm temperate to subtropical regions of the Northern Hemisphere (see Figs. 4.17 and 4.18). It can be found in Africa, South America, Europe, and Asia. It is a perennial herb, growing up to

Figure 4.17   *Cyperus esculentus* L. *(Photo courtesy of Rolv Hjelmstad [www.rolv.no/bilder/galleri/medplant/ cype_esc.htm].)*

90 cm high [138]. Tubers contain 20–36% oil. The oil from the tuber contains 18% saturated (palmitic acid and stearic acid) and 82% unsaturated (oleic acid and linoleic acid) fatty acids [138].

Main uses.   The tubers are edible and have high nutritive value. They contain 3–15% protein, 15–20% sugar, 20–25% starch, 4–14% cellulose, and trace amounts of natural resin. They are used in Spain to make a beverage named *horchata*, and also consumed fresh after soaking. In other countries, the tubers are used in sweetmeats or uncooked as a side dish. New products obtained can enhance the interest in this crop

Figure 4.18   *Cyperus esculentus* L. *(Photo courtesy of Peter Chen [www.cod.edu/people/faculty/ch enpe/PRAIRIE/2005_09_20/ Cyperus_esculentus.jpg].)*

as a source of dietary fiber in food technology, as a high-quality cooking/salad oil, as a source of starch, as an antioxidant-containing food, and so forth [139]. The oil extracted from yellow nutsedge can be used as food oil as well as for industrial purposes. Since the tubers contain 20–36% oil, the crop has been suggested as a potential oil crop for the production of biodiesel [138]. Preliminary tests using pure nutsedge oil as fuel in a diesel engine have indicated that the engine operated near its rated power [140]. Currently, it is being studied as an oil source for fuel production in Africa [53].

4.4   Used Frying Oils

Currently, world oil crop production is about 139,000,000 ton [141]. In particular, developing countries (97,370,185 ton) and developed countries (41,193,308 ton) are the largest producers, while least developed countries contribute 4,141,535 ton. Most of this oil is used for deep-frying processes, after which it becomes a disposal problem. Disposal methods often contaminate environmental water and contribute to world pollution. Due to high oxidative thermal stress, such waste frying fats should not be used for human food [142]. Also, since 2002, the EU has enforced a ban on feeding these mixtures to animals, because during frying many harmful compounds are formed, which could result in the return of harmful compounds back into the food chain through the animal meat [143].

Used oils can be recycled through conversion into soap by saponification and reused as lubricating oil or hydraulic fluid. Nevertheless, biofuel production seems to be the most attractive alternative for waste oil treatment. Certainly, it will not solve the energy problem, because only a small percentage of diesel demand can be supplied by this source [20], but it will decrease the dependence on fossil oil while reducing an environmental problem.

For economic reasons, used frying oil is an interesting feedstock for biodiesel production. In this sense, Nye et al. were the first to describe the transesterification of used frying oil using excess of alcohol under both acidic and basic conditions. The best result was obtained using methanol with catalysis by KOH [144]. The tests were carried out using frying margarine and partially hydrogenated soybean oil. The reaction was carried out at 50°C for 24 h, using methanol in a methanol–triglyceride molar ratio of 3.6:1 and 0.4% KOH. At the same time, Mittelbach et al. investigated the use of waste oils to produce biodiesel and found that the increase in the amount of polymers during heating of the oil is a good indicator for the suitability for biodiesel production [42]. They proposed a low-temperature process (40°C) under alkaline catalysis and excess of methanol [145]. Considering used olive oil, better results were also obtained using KOH and methanol instead of NaOH and ethanol, which decreases transesterification rates. The reaction was optimized at an ambient temperature, using 1.26% KOH and 12% methanol, and stirring for 1 min [40]. Some authors have optimized the reaction by using methanol (alcohol–waste oils molar ratios between 3.6 and 5.4) and 0.2–1% NaOH [146], or methanol (molar ratios in the range of 1:74 to 1:245) and acid catalyst (sulfuric acid) [147]. Al-Widyan and Al-Shyoukh have performed waste palm oil transesterification under various conditions. The best process combination was 2.25 M H_2SO_4 with 100% excess ethanol in about 3 h of reaction time [148].

Several parameters (e.g., heating conditions, FFA composition, and water content) can influence conversion from waste oils into biodiesel. Mittelbach et al. have found that heating over a long period led to a significantly higher FFA content, which can reach values up to 10% and have detrimental effects during the transesterification process. Nevertheless, in most cases, simple heating and filtering of solid impurities is sufficient for further transesterification [20]. The methyl and ethyl esters of fatty acids obtained by alcoholysis of triglycerides seem to be excellent fuels [5]. Anggraini found that it was also important to keep the water content of used cooking oils as low as possible [149]. Dorado et al. have compared biofuels from waste vegetable oils from several countries (different FFA composition) including Brazil, Spain, and Germany. The transesterification process was carried out in two steps, using a stoichiometric amount of methanol and the necessary amount of KOH, supplemented with the exact amount of KOH to neutralized acidity. Both reactions were completed in 30 min [41]. Results revealed that to carry the reaction to completion, an FFA value lower than 3% is needed. The two-step transesterification process (without any costly purification step) was found to be an economic method for biofuel production using waste vegetable oils. To reduce FFA content, a two-step transesterification using 0.2% ferric sulfate and 1% KOH with methanol (mole ratio 10:1) was also developed [150]. Acid-catalyzed pretreatment to esterifiy the FFA before transesterification with an alkaline catalyst was also proposed [151]. This procedure can reduce the acid levels to less than 1%. Some authors have proposed a three-step process in a fixed-bed bioreactor with immobilized *Candida antarctica* lipase [152]. Brenneis et al. also developed a process involving *C. antarctica* through alcoholysis of waste fats, with excess of water. The optimum amount of water was found to be 80–10% of the amount of fat [153]. Chen et al. preferred the use of immobilized lipase Novozym-435 in transesterification of both waste oil and methyl acetate. However, they found that the reaction rate decreased with increasing water content [154].

Engine tests have been performed with biodiesel from different kinds of waste oils. Al-Widyan et al. tested several ester–diesel blends in a direct-injection diesel engine. Results indicated that the biodiesel burned more efficiently with less specific fuel consumption. Furthermore, 50% of the blends produced less CO and fewer unburned hydrocarbons than diesel [155]. Also, Mittelbach and Junek stated that it improves exhaust gas emissions, as compared to esters made from fresh oil [156]. However, despite the exhaust emission reduction, there are some discrepancies in terms of NO_x emission related to the process and raw material [1, 105, 157]. In general terms, most studies show a slight decrease in brake power output, besides an increase in specific fuel consumption [158, 159]. To solve this problem, Kegl and Hribernik have proposed to modify injection characteristics at different fuel temperatures [160].

Several authors have worked on related topics. Kato et al. have used ozone treatment to reduce the flash point of biodiesel from fish waste oil, resulting in easy combustibility [161]. The immiscibility of canola oil in methanol provides a mass-transfer challenge in the early stages of transesterification. To exploit this situation, Dubé et al. developed a two-phase membrane reactor. The reactor was particularly useful in removing unreacted oil [162].

4.5 Animal Fats

Bovine spongiform encephalopathy (BSE), commonly known as mad cow disease, is a fatal neurodegenerative disease of cattle. BSE has attracted wide attention because it can be transmitted to humans. Pathogenic prions are responsible for transmissible spongiform encephalopathies (TSE), and especially for the occurrence of a new variant of Creutzfeldt-Jakob disease (nvCJD), a human brain-wasting disease. Due to this problem, the specified risk material is burned under high temperatures to avoid any hazards for humans and animals. However, another possibility could be to consider this material as a source for producing biodiesel by transesterification. In fact, production of biodiesel from the risk material could represent a more economic usage than its combustion. Siedel et al. have found that almost every single step of the process leads to a significant reduction in the concentration of the pathogenic prion protein (PrP^{Sc}) in the main product and by-products. They concluded that biodiesel from materials with a high concentration of pathogenic prions can be considered safe [163]. Animal fats, such as tallow or lard, have been widely investigated as a source of biodiesel [164–169]. Muniyappa et al. have found that transesterification of beef tallow produced a mixture of esters with a high concentration in saturated fatty acids, but with physical properties similar to esters of soybean oil [37]. Ma et al. found that 0.3% NaOH completed methanolysis of beef tallow in 15 min [170]. Some authors have found that absolute ethanol produced higher conversion and less viscosity than absolute methanol at 50°C, after 2 h [171]. Nebel and Mittelbach have found n-hexane was the most suitable solvent for extraction of fat from meat and bone meal. The extracted material was converted into fatty acid methyl esters through a two-step process [172]. Lee et al. have performed a three-step transesterification to produce biodiesel from lard and restaurant grease. They found that a porous substance, such as silica gel, improved the conversion when more than 1 M methanol was used as reaction substrate [173]. Mbaraka et al. also synthesized propylsulfonic acid-functionalized mesoporous silica materials for methanol esterification of the FFA in beef tallow, as a pretreatment step for alkyl ester production [174].

Engine tests also showed a reduction in emission, except oxides of nitrogen that increased up to 11% for the yellow grease methyl ester [157]. Cold-flow properties of the fat-based fuels were found to be less desirable than those of soy-based biodiesel, with comparable lubricity and oxidative stability [175]. To solve this problem, Kazancev et al. blended up to 25% of pork lard methyl esters with other oil methyl esters and fossil diesel fuels. In this case, the CFPP showed a value of −5°C. In winter, only up to 5% of esters can be added to the fuel. Depressant Viscoplex 10-35 with an optimal dose of 5000 mg/kg was found to be the most effective additive to improve the cold properties [101].

4.6   Future Lines

Research in most of the nonedible oil crops previously mentioned has been insufficient. To determine the viability of their use as a source of biodiesel and to optimize the transesterification as well as engine performance, more research is needed. But, there are also other nonedible and low-cost edible oily crops and trees that could be exploited for biodiesel production. Amongst them, allanblackia, bitter almond, chaulmoogra, papaya, sal, tung, and ucuuba produce oils that hold immense potential to be used as a raw material for producing biodiesel. Most of them grow in underdeveloped and developing countries, where governments may consider providing support to the activities related to collection of seeds, production of oil, production of biodiesel, and its utilization for cleaner environment. Hence, to facilitate its integration, a legal framework should be legislated to enforce regulations on biodiesel. Biodiesel should be seriously considered as a potential source of energy, particularly in underdeveloped and developing countries with very tight foreign exchange positions and insufficient availability of traditional fuels.

4.6.1 Allanblackia oil

Crop description. *Allanblackia stuhlmannii* and *A. floribunda*—commonly known as allanblackia, mkanyi fat, bouandjo, and kagne butter (see Fig. 4.19)—belong to the family Guttiferae and grow in tropical areas, mainly in East Africa, Congo, and Cameroons. A high content of hard white fat (60–80%) can be extracted from the seed kernels of the trees. Allanblackia fats consist almost entirely of stearic acid (52–58%), oleic acid (39–45%), and palmitic acid (2–3%) [87]. Allanblackia has received considerable attention, based on its fat composition rather than its commercial importance [77].

Main uses.   The use of the fat in soap manufacture has been suggested [176]. The timber is suitable for use under damp conditions. The pounded bark is used for medicinal purposes [177]. No references about its use as a biodiesel source have been found so far.

4.6.2 Bitter almond oil

Crop description. *Prunus communis, P. americana,* and *P. amygdalus*— commonly known as almond, amandier, mandelbaum, almendro, and mandorlo (see Fig. 4.20)—belong to the family Rosaceae and grow in temperate Mediterranean areas. Major producing countries are Italy, Spain, Morocco, France, Greece, and Iran. The almond tree grows to a height of 3–8 m. Many varieties of almonds are grown, but they can be

Figure 4.19 *Allanblackia stuhl-mannii. (Photo courtesy of Josina Kimottho (ICRAF) [www.worldagroforestry.org/Sites/TreeDBS/aft/imageSearch.asp].)*

Figure 4.20 *Prunus communis. (Photo courtesy of Gernot Katzer [www.uni-graz.at/~katzer/pictures/prun_09.jpg].)*

broadly divided into two types: bitter and sweet. Bitter almonds contain amygdalin and an enzyme that causes its hydrolysis to glucose, benzaldehyde, and hydrocyanic acid, making the fruit nonedible. The bitter almond oil yield is around 40–45%, and sometimes as low as 20% [77, 178]. Major fatty acid composition of oil includes palmitic acid (7.5%), stearic acid (1.8%), oleic acid (66.4%), and linoleic acid (23.5%) [178].

Main uses.   Bitter almond press cake cannot be used for feed due to its toxic components [179]. They are pressed at low temperatures, generally at about 30°C, to prevent destruction of the hydrolytic enzyme. The press cake is then used for production of bitter almond oil [77]. Despite the oil content and fatty acid composition, no references about the use of bitter almond oil as a raw material to produce biodiesel have been found so far.

4.6.3 Chaulmoogra oil

Crop description.   *Taraktogenos kurzii, Hydnocarpus wightiana, Oncoba echinata* (West Africa), and *Carpotroche brasiliensis* (Brazil)—commonly known as chaulmoogra, chaulmugra, maroti, hydnocarpus, and gorli seed—belong to the family Flacourtiaceae and grow in India, Sri Lanka, Burma, Bangladesh, Nigeria, and Uganda (see Fig. 4.21). The trees

Figure 4.21   Chaulmoogra leaves.
(*Photo courtesy of Prof. Gerald D. Carr [www.botany.hawaii.edu/faculty/carr/flacourti.htm].*)

grow to a height of 12–15 m. The kernels make up 60–70% of the seed weight and contain 63% of pale-yellow oil. The oil is unusual in that it is not made up of straight-chain fatty acids but acids with a cyclic group at the end of the chain [77].

Main uses.   Chaulmoogra oil has been used for thousands of years in the treatment of leprosy. However, it has now been replaced by modern drugs. The expeller cake is a useful manure and is reported to ward off ants and other insect pests. It cannot be used for animal feed due to its toxicity. The oil has been highly active against fungal plant pathogens including *Aspergillus niger* and *Rhizopus nigricans*. There may be a wide scope of integrating the pharmaceutical industries based on chaulmoogra, with the fuel and energy industries dealing with production of petroleum hydrocarbons, such as biodiesel [180].

4.6.4.   Papaya oil

Crop description.   *Carica papaya* L. (see Fig. 4.22)—commonly known as papaya, pawpaw, melon tree, papayier, lechosa, or mamon—belongs to the family Caricaceae and grows in tropical to subtropical areas. Native to South America, now the crop is widely distributed throughout the tropics. Papaya is a short-lived rapidly growing plant (not a true tree) having no lignified tissues. The seeds contain 25–29% oil [77, 179]. The oil contains mainly unsaturated fatty acids, around 70.7%, and may contain toxic components that make it unusable in human foods [75]. Fatty acid composition of the oil includes oleic acid (79.1%) and palmitic acid (16.6%) [179].

Main uses.   Papaya is mainly used as fresh fruit, and for the production of drinks, jams, and so forth. In some places, the seeds are used for treatment against worms [181]. The green fruit is also a commercial source

Figure 4.22   *Carica papaya* L. (*Photo courtesy of Barbara Simonsohn [www.barbara-simonsohn.de/papaya.htm].*)

of the proteolytic enzymes papain and chymopapain—the former finding use in a wide range of industries, particularly brewing for haze removal, and the latter in medicine. Oil extraction from the seeds could improve the viability of the industry in countries where papaya is cultivated for papain production and processing. The seeds constitute around 22% of the waste from papaya puree plants [182]. No references about its use as a biodiesel source have been found so far.

4.6.5 Sal oil

Crop description.   *Shorea robusta* Gaertn. f.—commonly known as sal, shal, saragi, sakhu, sakher, shaal, ral, gugal, mara, sagua, salwa, sakwa, kandar, and kung—is a large tree belonging to the family Dipterocarpaceae (see Fig. 4.23). The tree is native to southern Asia, ranging south of the Himalayas, from Myanmar in the east to India, Bangladesh, and Nepal. It grows in dry tropical forests, in a well-drained, moist, sandy loam soil. This tree can attain heights up to 35 m. The seeds of sal are an important source of edible oil. The seed contains around 20% of oil [183, 184].

Main uses.   Although sal is a highly valued timber species, it is also used for house construction, and as poles, agriculture implements, fuelwood, fencing, leaves for cups and plates, and compost [185]. The oil is used for lighting and cooking purposes, and as a substitute for

Figure 4.23 *Shorea robusta* Gaertn. f. *(Photo courtesy of Dr. Mike Kuhns [http://extension.usu.edu/forestry/ UtahForests/TreeID/Assets/ Images/sal-1.3.jpg].)*

cocoa butter in the manufacture of chocolates. It is suitable for soap making after blending with other softer oils. The oil cakes that remain after oil extraction contain 10–12% protein and about 50% starch, and are used as cattle and poultry feed. However, the oil cake contains 5–14% tannin; consequently, not more than 20% is concentrated for cattle without detrimental effects. As the protein remains completely undigested, the oil cake yields energy only. Sal resin is burned as incense in Hindu ceremonies. It is also used for varnishes, for hardening softer waxes for use in the manufacture of shoe polishes, and as cementing material for plywood, asbestos sheets, and so forth. The resin is used in an indigenous system of medicine as an astringent and detergent [184]. No references about its use as a biodiesel source have been found so far.

4.6.6 Tung oil

Crop description. *Aleurites fordii (Vernicia fordii)* and *A. montana*—commonly known as the tung tree, Chinese wood, Abrasin, and Mu (see Fig. 4.24)—belong to the family Euphorbiaceae and grow well in cold climates, but will survive in subtropical conditions (*A. fordii*). *A. montana* prefers a tropical climate. Major producers are China, Argentina, Paraguay, Brazil, and the United States. The nut of this deciduous tree contains an oil-rich kernel. The oil content of the air-dried fruit lies between 15% and 20% [77]. Major fatty acid composition of oil includes

Figure 4.24   *Vernicia fordii. (Photo courtesy of Dr. Alvin R. Diamond [http://spectrum. troy.edu/~diamond/PIKEFLORA.htm].)*

palmitic acid (5.5%), oleic acid (4.0%), linoleic acid (8.5%), and eleostearic acid (82%) [77].

Main uses.   Tung oil is used in paints, varnishes, and so forth. It is also used in the production of linoleum, resins, and chemical coatings. It has been used in motor fuels in China [77]. The seed cake after oil extraction is used as a fertilizer and cannot be used for animal feed as it contains a toxic protein [75]. No references about its use as a raw material to produce biodiesel have been found to date.

4.6.7   Ucuuba oil

Crop description.   *Virola surinamensis* and *V. sebifera* (see Fig. 4.25)—commonly known as ucuhuba, ucuiba, ucuba, muscadier porte-suif, and yayamadou—belong to the family Myristicaceae and grow in tropical swampy forests. Major producing countries are Brazil, Costa Rica, Ecuador, French Guiana, and Guyana. A typical tree is of medium height and can produce 60–90 L of oil each year. The seeds contain 65–76% oil. The yellow-brown aromatic oils from both varieties are very similar. Other related species, such as *V. otoba*, which grows in Colombia and Peru, yield a fat similar to ucuuba, which is known as otoba butter or American nutmeg butter. Major fatty acids present in the oil are lauric

Figure 4.25   Ucuuba tree. (*Photo courtesy of Eugênio Arantes de Melo [www.arvores.brasil.nom.br/].*)

acid (15–17.6%), myristic acid (72.9–73.3%), palmitic acid (4.4–5%), and oleic acid (5.1–6.3%) [77, 87].

Main uses.   This fat has been used traditionally in candle manufacture. The fat and pulverized kernels find use in traditional medicines. The tree has been proposed as a potential source of isopropyl myristate, which is used in cosmetic manufacture [186]. However, no references related to its use as a raw material to produce biodiesel have been found to date.

Acknowledgments

My sincere thanks to the following people and organizations for their generosity in letting me use their photos: Dr. Kazuo Yamasaki (Teikyo Heisei University, Japan), Abdulrahman Alsirhan (www.alsirhan.com), Eric Winder (Biological Sciences, Michigan Technological University), Jack Bacheler (Department of Entomology, North Carolina State University), Dr. Alvin R. Diamond (Department of Biological and Environmental Sciences, Troy University), Piet Van Wyk and EcoPort, Food and Agricultural Organization of the United Nations, Antoine van den Bos (Botanypictures), Forest and Kim Starr (USGS), Dr. Davison Shillingford (Dominica Academy of Arts and Sciences), Prof. Arne Anderberg (Swedish Museum of Natural History), Rolv Hjelmstad (Urtekilden), Peter Chen (College of DuPage), Josina Kimottho (ICRAF), Gernot Katzer (University of Graz), Prof. Gerald D. Carr (University of Hawaii, Botany Department), Barbara Simonsohn, Dr. Mike Kuhns (Utah State University), and Eugênio Arantes de Melo (Árvores do Brasil).

References

1. M. P. Dorado, et al. Exhaust emissions of a diesel engine fueled with transesterified waste vegetable oil, *Fuel* **82**(11), 1311–1315, 2003.
2. C. Peterson and D. Reece. Emissions characteristics of ethyl and methyl ester of rapeseed oil compared with low sulfur diesel control fuel in a chassis dynamometer test of a pickup truck, *Transactions of the ASAE* **39**(3), 805–816, 1996.
3. D. Y. Z. Chang, et al. Fuel properties and emissions of soybean oil esters as diesel fuel, *JAOCS* **73**(11), 1549–1555, 1996.
4. L. G. Schumacher, et al. Heavy-duty engine exhaust emission tests using methyl ester soybean oil/diesel fuel blends, *Bioresource Technology* **57**(1), 31–36, 1996.
5. M. Mittelbach and P. Tritthart. Diesel fuel derived from vegetable oils, III. Emission tests using methyl esters of used frying oil, *JAOCS* **65**(7), 1185–1187, 1988.
6. C. A. Sharp. Exhaust emissions and performance of diesel engines with biodiesel fuels, In: *Biodiesel Environmental Workshop*, 1998, Southwest Research Institute, Washington DC.
7. M. Mittelbach, P. Tritthart, and H. Junek. Diesel fuel derived from vegetable oils, II: Emission tests using rape oil methyl ester, *Energy in Agriculture* **4**, 207–215, 1985.
8. S. M. Geyer, M. J. Jacobus, and S. S. Lestz. Comparison of diesel engine performance and emissions from neat and transesterified vegetable oils, *transactions of the ASAE* **27**(2), 375–384, 1984.
9. R. L. McCormick, M. S. Graboski, and T. L. Alleman. *Health Related Emissions from Biodiesel*, 1999, Colorado Institute for Fuels and Engine Research, Colorado School of Mines, Colorado.
10. N. N. Clark and D. W. Lyons. Class 8 truck emissions testing: Effects of test cycles and data on biodiesel operation, *Transactions of the ASAE* **42**(5), 1211–1219, 1999.
11. M. P. Dorado, et al. Testing waste olive oil methyl ester as a fuel in a diesel engine, *Energy & Fuels* **17**(6), 1560–1565, 2003.
12. C. Adams, et al. Investigation of soybean oil as a diesel fuel extender: Endurance tests, *JAOCS* **60**(8), 1574–1579, 1983.
13. J. Einfalt and C. E. Goering. Methyl soyate as a fuel in a diesel tractor, *Transactions of the ASAE* **28**(1), 70–74, 1985.
14. J. Fuls, C. S. Hawkins, and F. J. C. Hugo. Tractor engine performance on sunflower oil fuel, *Journal of Agricultural Engineering Research* **30**, 29–35, 1984.
15. K. J. Harrington. Chemical and physical properties of vegetable oil esters and their effect on diesel fuel performance, *Biomass* **9**, 1–17, 1986.
16. C. L. Peterson, D. L. Reece, R. Cruz, and J. Thompson. A comparison of ethyl and methyl esters of vegetable oil as diesel fuel substitutes: Liquid fuels from renewable resources, In: *Proceedings of an Alternative Energy Conference,* American Society of Agricultural Engineers, 99–110, 1992, USA.
17. F. Karaosmanoglu, A. Akdag, and K. B. Cigizoglu. Biodiesel from rapeseed oil of Turkish origin as an alternative fuel, *Applied Biochemistry and Biotechnology* **61**(3), 251–265, 1996.
18. K. R. Kaufman and M. Ziejewski. Sunflower methyl esters for direct injected diesel engines, *Transactions of the ASAE* **27**, 1626–1633, 1984.
19. R. Loreto, L. Martínez, and D. Rivas. Performance of a diesel engine working with mixtures of diesel–coconut oil, In: *XI Congreso Nacional de Ingeniería Mecánica,* 1994, Valencia, Spain.
20. M. Mittelbach, B. Pokits, and A. Silberholz. Diesel fuels derived from vegetable oils, IV: Production and fuel properties of fatty acid methyl esters from used frying oil, In: *Liquid Fuels from Renewable Resources. Proceedings of an Alternative Energy Conference*, 1992, Nashville, Tennessee.
21. F. Moreno, M. Muñoz, and J. Morea-Roy. Sunflower methyl ester as a fuel for automobile diesel engines, *Transactions of the ASAE* **42**(5), 1181–1185, 1999.
22. R. A. Niehaus, et al, Cracked soybean oil as a fuel for a diesel engine, *Transactions of the ASAE* **29**(3), 683–689, 1986.
23. C. L. Peterson. Vegetable oil as a diesel fuel: Status and research priorities, *Transactions of the ASAE* **29**(5), 1413–1422, 1986.
24. C. L. Peterson. Production and testing of ethyl and methyl esters, In: *Biodiesel Recipes from New Oil*, 1994, National Biodiesel Board, University of Idaho, Moscow, USA.

25. C. L. Peterson and D. Reece. *On-Road Testing of Biodiesel—A Report of Past Research Activities,* 1997, Department of Agricultural Engineering, University of Idaho, Moscow, USA.
26. G. Pischinger, A. M. Falcon, and R. W. Siekmann. Soybean ester as alternative diesel fuel tested in DI engine powered Volkswagen trucks, In: *Vegetable Oils as Diesel Fuel, Seminar III, Agricultural Reviews and Manuals,* 1983, Peoria, Ilinois.
27. G. Pischinger, et al. Results of engine and vehicle tests with methyl esters of plant oils as alternative diesel fuels, In: *International First Symposium on Alcohol Fuel Technology,* 1982, Auckland, New Zealand.
28. R. W. Pryor, et al. Soybean oil fuel in a small diesel engine, *Transactions of the ASAE* **26**(2), 333–337, 1983.
29. D. L. Purcell, et al. Transient testing of soy methyl ester fuels in an indirect injection, compression ignition engine, *JAOCS* **73**(3), 381–388, 1996.
30. M. P. Dorado, et al. An approach to the economics of two vegetable oil-based biofuels in Spain, *Renewable Energy* **31**, 1231–1237, 2006.
31. M. Bender. Economic feasibility review for community-scale farmer cooperatives for biodiesel, *Bioresource Technology* **70**, 81–87, 1999.
32. D. L. Van Dyne, J. A. Weber, and C. H. Braschler. Macroeconomic effects of a community-based biodiesel production system, *Bioresource Technology* **56**(1), 1–6, 1996.
33. M. W. Formo. Ester reactions of fatty materials, *JAOCS* **31**, 548–559, 1954.
34. B. Freedman, E. H. Pryde, and T. L. Mounts, Variables affecting the yields of fatty esters from transesterified vegetable oils, *JAOCS* **61**(10), 1638–1643, 1984.
35. C. S. Hawkins and J. Fuls. Comparative combustion studies on various plant oil esters and the long term effects of an ethyl ester on a compression ignition engine, In: *Proceedings of the International Conference on Plant and Vegetable Oils as Fuels,* ASAE, 184–97, 1982, St. Joseph, Michigan.
36. A. Isigigür, F. Karaosmanoglu, and H.A. Aksoy. Methyl ester from safflower seed oil of Turkish origin as a biofuel for diesel engines, *Applied Biochemistry and Biotechnology* **45**(46), 103–112, 1994.
37. P. R. Muniyappa, S. C. Brammer, and H. Noureddini. Improved conversion of plant oils and animal fats into biodiesel and co-product, *Bioresource Technology* **56**(1), 19–24, 1996.
38. D. N. Zheng and M.A. Hanna. Preparation and properties of methyl esters of beef tallow, *Bioresource Technology* **57**(2), 137–142, 1996.
39. M. P. Dorado, et al. Optimization of alkali-catalyzed transesterification of *Brassica carinata* oil for biodiesel production, *Energy & Fuels* **18**(1), 77–83, 2004.
40. M. P. Dorado, et al. Kinetic parameters affecting the alkali-catalyzed transesterification process of used olive oil, *Energy & Fuels* **18**(5), 1457–1462, 2004.
41. M. P. Dorado, et al. An alkali-catalized transesterification process for high free fatty acid feedstocks, *Transactions of the ASAE* **45**(3), 525–529, 2002.
42. M. Mittelbach, and H. Enzelsberger. Transesterification of heated rapeseed oil for extending diesel fuel, *JAOCS* **76**(5), 545–550, 1999.
43. A. A. Anggraini, R. Krause, and H.-P. Löhrlein. Altöle und-fette verbrennen, *Energie* **3**, 140–141, 1997.
44. T. Krawczyk. Biodiesel, In: *International News on Fats, Oils and Related Materials,* 1996, American Oil Chemists Society Press, Champaign, Illinois, p. 801.
45. J. Connemann and J. Fischer. Biodiesel in Europe 1998: Biodiesel processing technologies, In: *International Liquid Biofuels Congress,* 1998, Curitiba, Parana, Brazil.
46. Y. Zhang, et al. Biodiesel production from waste cooking oil: 2. Economic assessment and sensitivity analysis, *Bioresource Technology* **90**(3), 229–240, 2003.
47. K. A. Subramanian, et al. Utilization of liquid biofuels in automotive diesel engines: An Indian perspective, *Biomass and Bioenergy* **29**, 65–72, 2005.
48. M. T. Makwana, J. S. Patolia, and E. R. R. Iyengar. *Salvadora* plant species suitable for saline coastal wasteland, *Transactions of Indian Society of Desert Technology* **2**, 121–131, 1988.
49. A. U. Khan. History of decline and present status of natural tropical thorn forest in Punjab, *Biological Conservation* **67**, 205–210, 1994.
50. A. U. Khan. Appraisal of ethno-ecological incentives to promote conservation of *Salvadora oleoides* decne: The case for creating a resource area, *Biological Conservation* **75**, 187–190, 1996.

51. A. K. Bhatnagar, et al. HFRR studies on methyl esters of nonedible vegetable oils, *Energy & Fuels* **20**(3), 1341–1344, 2006.
52. L. Roth and K. Kormann. *Ölpflanzen Pflanzenöle, Fette Wachse Fettsäuren Botanik Inhaltstoffe Analytick*, Landsberg/Lech, Germany: Ecomed Verlagsgesellschaft, 2000.
53. M. Mittelbach and C. Remschmidt. *Biodiesel: The Comprehensive Handbook*, Graz, Austria: Martin Mittelbach, 2004, p. 170.
54. C. Devendra and G.V. Raghavan. Agricultural by-products in South East Asia: Availability, utilization and potential value, *World Review of Animal Production* **14**(4), 11–27, 1978.
55. D. de Oliveira, et al. Optimization of alkaline transesterification of soybean oil and castor oil for biodiesel production, *Applied Biochemistry and Biotechnology* **121**, 553–560, 2005.
56. F. F. Fagundes, et al. Lipase catalyzed synthesis of biodiesel from castor oil in organic solvent *Journal of Biotechnology* **118**(1), 166–169, 2005.
57. J. S. Yang, et al. Enzymatic methanolysis of castor oil for the synthesis of methyl ricinoleate in a solvent-free medium, *Journal of Microbiology and Biothechnology* **15**(6), 1183–1188, 2005.
58. S. M. P. Meneghetti, et al. Ethanolysis of castor and cottonseed oil: A systematic study using classical catalysts, *JAOCS* **83**(9), 819–822, 2006.
59. S. M. P. Meneghetti, et al. Biodiesel from castor oil: A comparison of ethanolysis versus methanolysis, *Energy & Fuels* **20**(5), 2262–2265, 2006.
60. J. Cvengros, J. Paligova, and Z. Cvengrosova. Properties of alkyl esters base on castor oil, *European Journal of Lipid Science Technology* **108**, 629–635, 2006.
61. R. H. M. Langer and G. D. Hill. *Agricultural Plants*, Second ed., Cambridge: Cambridge University Press, 1991.
62. L. Miles. *Focus on Cotton*, Hove, East Sussex, UK: Wayland (Publishers) Ltd., 1986.
63. J. M. Munro. *Cotton*, Second ed., Tropical Agriculture Series, Harlow: Longman Scientific and Technical, 1987.
64. D. K. Salunkhe, et al. *World Oilseeds: Chemistry Technology and Utilisation*, New York: Van Nostrand Reinhold, 1992.
65. Ö. Köse, M. Tüter, and H. A. Aksoy. Immobilized *Candida antarctica* lipase-catalyzed alcoholysis of cotton seed oil in a solvent-free medium, *Bioresource Technology* **83**, 125–129, 2002.
66. T. W. Tan, K. L. Nie, and F. Wang. Production of biodiesel by immobilized *Candida* sp. lipase at high water content, *Applied Biochemistry and Biotechnology* **128**(2), 109–116, 2006.
67. D. Royon, et al. Enzymatic production of biodiesel from cotton seed oil using *t*-butanol as a solvent, *Bioresource Technology* **98**, 648–653, 2007.
68. N. G. Siatis, et al. Improvement of biodiesel production based on the application of ultrasound: Monitoring of the procedure by FTIR spectroscopy, *JAOCS* **83**(1), 53–57, 2006.
69. I. Kaliangilee and D. F. Grabe. Seed maturation in *Cuphea*, *Journal of Seed Technology* **12**(2), 107–113, 1988.
70. S. A. Graham. *Cuphea*: A new plant source of medium chain fatty acids, *CRC Critical Reviews in Food Science and Nutrition* **28**(2), 139–173, 1989.
71. D. P. Geller, J. W. Goodrum, and S. J. Knapp. Fuel properties of oil from genetically altered *Cuphea viscosissima*, *Industrial Crops and Products* **9**, 85–91, 1999.
72. D. P. Geller and J. W. Goodrum. Rheology of vegetable oil analogs and triglycerides, *JAOCS* **77**(2), 111–114, 2000.
73. D. P. Geller, J. W. Goodrum, and C. C. Campbell. Rapid screening of biologically modified vegetable oils for fuel performance, *Transactions of the ASAE* **42**(4), 859–862, 1999.
74. D. P. Geller, J. W. Goodrum, and E. A. Siesel. Atomization of short-chain triglycerides and a low molecular weight vegetable oil analogue in DI diesel engines, *Transactions of the ASAE* **46**(4), 955–958, 2003.
75. N. J. Godin and P. C. Spensley. Oils and oilseeds, In: *Crop and Products Digest No. 1*, Tropical Product Institute, London, p. 170, 1971.
76. E. Winkler, et al. Enzyme-supported oil extraction from *Jatropha curcas* seeds, *Applied Biochemistry and Biotechnology* **63**(5), 449–456, 1997.

77. E. W. Eckey. *Vegetable Fats and Oils*, New York: Reinhold Publishing Corp., p. 559, 1954.
78. M. M. Azam, A. Waris, and N. M. Nahar. Prospects and potential of fatty acid methyl esters of some nontraditional seed oils for use as biodiesel in India, *Biomass and Bioenergy* **29**, 293–302, 2005.
79. N. Kumar and P. B. Sharma. *Jatropha curcas*—A sustainable source for production of biodiesel, *Journal of Scientific & Industrial Research* **64**(11), 883–889, 2005.
80. P. Vasudevan, S. Sharma, and A. Kumar. Liquid fuel from biomass: An overview, *Journal of Scientific & Industrial Research* **64**(11), 822–831, 2005.
81. N. Foidl, et al. *Jatropha curcas* L. as a source for the production of biofuel in Nicaragua, *Bioresource Technology* **58**, 77–82, 1996.
82. S. Shah, S. Sharma, and M. N. Gupta. Enzymatic transesterification for biodiesel production, *Indian Journal of Biochemistry & Biophysics* **40**(6), 392–399, 2003.
83. S. Shah, S. Sharma, and M. N. Gupta. Biodiesel preparation by lipase catalyzed transesterification of *Jatropha* oil, *Energy & Fuels* **18**(1), 154–159, 2004.
84. M. K. Modi, et al. Lipase-mediated transformation of vegetable oils into biodiesel using propan-2-ol as acyl acceptor, *Biotechnology Letters* **28**(9), 637–640, 2006.
85. H. Zhu, et al. Preparation of biodiesel catalyzed by solid super base of calcium oxide and its refiining process, *Chinese Journal of Catalysis* **27**(5), 391–396, 2006.
86. B. Mandal, S. G. Majumdar, and C. R. Maity. Chemical and nutritional evaluation of *Pongamia glabra* oil and *Acacia auriculaeformis* oil, *JAOCS* **61**(9), 1447–1449, 1984.
87. T. P. Hilditch. *The Chemical Composition of Natural Fats*, Second ed., London: Chapman & Hall Ltd., p. 188, 1947.
88. N. V. Bringi. *Nontraditional oilseeds and oils in India*, New Delhi, India: Oxford & IBH Publishing Co. Pvt. Ltd., p. 254, 1987.
89. A. G. Phadatare. *Performance and Emission Study of Power Tiller Engine Using Biodiesel*, Agricultural and Food Engineering Department, 2003, IIT Kharagpur, Kharagpur, India.
90. L. C. Meher, S. N. Naik, and L. M. Das. Methanolysis of *Pongamia pinnata* (karanja) oil for production of biodiesel, *Journal of Scientific & Industrial Research* **63**(11), 913–918, 2004.
91. Vivek and A. K. Gupta. Biodiesel production from Karanja oil, *Journal of Scientific & Industrial Research* **63**(1), 39–47, 2004.
92. H. Raheman, and A. G. Phadatare. Diesel engine emissions and performance from blends of karanja methyl ester and diesel, *Biomass and Bioenergy* **27**, 393–397, 2004.
93. L. C. Meher, et al. Transesterification of karanja *(Pongamia pinnata)* oil by solid basic catallysts, *European Journal of Lipid Science Technology* **108**, 389–397, 2006.
94. B. K. De and D. K. Bhattacharyya. Biodiesel from minor vegetable oils like karanja oil and nahor oil, *Fett/Lipid* **10**, 404–406, 1999.
95. R. K. Singh, A. K. Kumar, and S. Sethi. Preparation of karanja oil methyl ester, In: *Offshore World*, Mumbai, India: Jasubhai Media Pvt. Ltd., pp. 1–7, 2006 (April–May).
96. Y. Coskuner and E. Karababa. Some physical properties of flaxseed (*Linun usitatissimum* L.), *Journal of Food Engineering* **78**, 1067–1073, 2007.
97. P. Ody. *The Herb Society's Complete Medicinal Herbal*, London: Dorling Kindersley, 1993.
98. D. R. Berglund. Flax: New uses and demands, In: *Trends in New Crops and New Uses*, Janik, J. and Whipkey, A. (Eds.), Alexandria, VA: ASHS Press, pp. 358–360, 2002.
99. E. S. Oplinger, et al. Flax, In: *Alternative Field Crops Manual*, 1989 [cited; available from: www.hort.purdue.edu/newcrop/afcm/flax.html].
100. X. Lang, et al. Preparation and characterization of bio-diesels from various bio-oils, *Bioresource Technology* **80**, 53–62, 2001.
101. K. Kazancev, et al. Cold flow properties of fuel mixtures containing biodiesel derived from animal fatty waste, *European Journal of Lipid Science Technology* **108**, 753–758, 2006.
102. A. K. Agarwal, J. Bijwe, and L. M. Das. Wear assessment in a biodiesel fueled compression ignition engine, *Journal of Engineering for Gas Turbines and Power* **125**(3), 820–826, 2003.

103. A. K. Agarwal. Experimental investigations of the effect of biodiesel utilization on lubricating oil tribology in diesel engines, *Proceedings of the Institution of Mechanical Engineers* **219**(D5), 703–713, 2005.

104. E. Sendzikiene, V. Makarevicience, and P. Janulis. Oxidation stability of biodiesel fuel produced from fatty wastes, *Polish Journal of Environmental Studies* **14**(3), 335–339, 2005.

105. S. Lebedevas, et al. Use of waste fats of animal and vegetable origin for the production of biodiesel fuel: Quality, motor properties, and emissions of harmful components, *Energy & Fuels* **20**(5), 2274–2280, 2006.

106. NOVODBOARD. *Mahua* [cited 2006; available from: www.novodboard.com/ Madhua.pdf].

107. S. V. Ghadge and H. Raheman. Biodiesel production from mahua *(Madhuca indica)* oil having high free fatty acids, *Biomass and Bioenergy* 28, 601–605, 2005.

108. S. Puhan, et al. Performance and emission study of Mahua oil *(Madhuca indica* oil) ethyl ester in a 4-stroke natural aspirated direct injection diesel engine, *Renewable Energy* **30**, 1269–1278, 2005.

109. S. Puhan, et al. Mahua oil *(Madhuca indica* seed oil) methyl ester as biodiesel-preparation and emission characteristics, *Biomass and Bioenergy* **28**, 87–93, 2005.

110. S. V. Ghadge and H. Raheman. Process optimization for biodiesel production from mahua *(Madhuca indica)* oil using response surface methodology, *Bioresource Technology* **97**, 379–384, 2006.

111. S. Puhan, et al. Mahua *(Madhuca indica)* seed oil: A source of renewable energy in India, *Journal of Scientific & Industrial Research* **64**(11), 890–896, 2005.

112. C. Morel, et al. New xanthones from *Calophyllum caledonicum, Journal of Natural Products* **63**, 1471–1474, 2000.

113. J. Fournet. *Flore illustrée des Phanérogames de Guadeloupe et de Martinique*, Gondwana Editions, Vol. 1, Paris: INRA, 2002.

114. J. Hemavathy and J. V. Prabhakar. Lipid composition of *Calophyllum inophyllum* kernel, *JAOCS* **67**, 955–957, 1990.

115. S. Crane, et al. Composition of fatty acids triacylglycerols and unsaponifiable matter in *Calophyllum calaba* L. oil from Guadeloupe, *Phytochemistry* **66**, 1825–1831, 2005.

116. F. Cottiglia, et al. New chromanone acids with antibacterial activity from *Calophyllum brasiliense, Journal of Natural Products* **67**, 537–541, 2004.

117. N. Nabi, S. Akhter, and M. Z. Shahadat. Improvement of engine emissions with conventional diesel fuel and diesel–biodiesel blends, *Bioresource Technology* **97**, 372–378, 2006.

118. T. P. Hilditch. Variations in composition of some linolenic-rich seed oil, *Journal of the Science of Food and Agriculture* **2**, 543–547, 1951.

119. C. F. Reed. *Information Summaries on 1000 Economic Plants*, 1976, USDA.

120. O. E. Ikwuagwu, I. C. Ononogbu, and O. U. Njoku. Production of biodiesel using rubber [*Hevea brasiliensis* (Kunth. Muell.)] seed oil, *Industrial Crops and Products* **12**, 57–62, 2000.

121. A. S. Ramadhas, C. Muraleedharan, and S. Jayaraj. Performance and emission evaluation of a diesel engine fueled with methyl esters of rubber seed oil, *Renewable Energy* **30**, 1789–1800, 2005.

122. F. R. Abreu, et al. Utilization of metal complexes as catalysts in the transesterification of Brazilian vegetable oils with different alcohols. *Journal of Molecular Catalysis A: Chemical* **209**, 29–33, 2004.

123. A. L. Graciani, et al. *Brassica carinata, Cynara cardunculus* and *Sinapis alba* as three energetic crops, In: *1st World Conference and Exhibition on Biomass for Energy and Industry*, London: James & James (Science Publishers) Ltd., pp. 1560–1561, 2001.

124. J. M. Encinar, et al. Preparation and properties of biodiesel form *Cynara cardunculus* L. oil, *Industrial & Engineering Chemistry Research* **38**(8), 2927–2931, 1999.

125. J. M. Encinar, et al. Biodiesel fuel from vegetable oils: Transesterification of *Cynara cardunculus* L. oils with ethanol, *Energy & Fuels* **16**(2), 443–450, 2002.

126. M. Lapuerta, et al. Diesel emissions from biofuels derived from Spanish potential vegetable oils, *Fuel* **84**, 773–780, 2005.

127. D. S. Kimber and D. I. McGregor. The species and their origin, cultivation and world production, In: *Brassica Oilseeds: Production and Utilization*, Wallingford, UK: C. International, pp. 1–7, 1995.
128. N. J. Mendham and P. A. Salisbury. Physiology: Crop development, growth and yield, In: *Brassica Oilseeds: Production and Utilization*, Wallingford, UK: C. International, pp. 11–64, 1995.
129. M. Cardone, et al. *Brassica carinata* as an alternative oil crop for the production of biodiesel in Italy: Agronomic evaluation, fuel production by transesterification and characterization, *Biomass and Bioenergy* **25**, 623–636, 2003.
130. T. Genet, M. T. Labuschangne, and A. Hugo. Genetic relationships among Ethiopian mustard genotypes based on oil content and fatty acid composition, *African Journal of Biotechnology* **4**(11), 1256–1268, 2005.
131. M. P. Dorado, E. Ballesteros, and F. J. Giménez. *Biocombustibles para motores diesel procedente de ésteres metílicos de aceite de Brassica carinata sin ácido erúcico*, 2005, University of Jaen, University of Cordoba, Spain.
132. M. Cardone, et al. *Brassica carinata* as an alternative oil crop for the production of biodiesel in Italy: Engine performance and regulated and unregulated exhaust emissions, *Environmental Science & Technology* **36**(21), 4656–4662, 2002.
133. J. T. Budin, W. M. Breene, and D. H. Putnam. Some compositional properties of camelina (*Camelina sativa* L. Crantz) seeds and oils, *JAOCS* **72**(3), 309–315, 1995.
134. J. Vollmann, et al. Improvement of *Camelina sativa*, an underexploited oilseed, In: *Progress in New Crops*, Janick, J. (Ed.), Alexandria, VA: ASHS Press, pp. 357–362, 1996.
135. A. Fröhlich and B. Rice. Evaluation of *Camelina sativa* oil as a feedstock for biodiesel production, *Industrial Crops and Products* **21**, 25–31, 2005.
136. G. Steinke, R. Kirchhoff, and K. D. Mukherjee. Lipase-catalyzed alcoholysis of crambe oil and camelina oil for the preparation of long-chain esters, *JAOCS* **77**(4), 361–366, 2000.
137. G. Steinke, S. Schonwiese, and K. D. Mukherjee. Alkali-catalyzed alcoholysis of crambe oil and camelina oil for the preparation of long-chain esters, *JAOCS* **77**(4), 367–371, 2000.
138. H. Y. Zhang, et al. Yellow nut-sedge (*Cyperus esculentus* L.) tuber oil as a fuel, *Industrial Crops and Products* **5**, 177–181, 1996.
139. B. Pascual, et al. Chufa *(Cyperus esculentus* L. var. *sativis Boeck.)*: An unconventional crop. Studies related to applications and cultivation, *Economic Botany* **54**(4), 439–448, 2000.
140. L. Nan, et al. Use of yellow nut-sedge oil as a substitute for diesel fuel, In: *ASAE Paper*, 1984, ASAE, St. Joseph, MI.
141. FAO. *World oilcrops production,* FAOSTAT, FAO Statistics Division, 2006 (available from: http://faostat.fao.org/site/408/DesktopDefault.aspx?PageID=408, 2005).
142. S. S. Chang and C. L. Peterson. Chemical reactions involved in the deep-fat frying of foods, *JAOCS* **55**, 718–727, 1978.
143. M. G. Kulkarni and A. K. Dalai. Waste cooking oil—An economical source for biodiesel: A review, *Industrial & Engineering Chemistry Research* **45**(9), 2901–2913, 2006.
144. M. J. Nye, et al. Conversion of used frying oil to diesel fuel by transesterification: Preliminary tests, *JAOCS* **60**(8), 1598–1601, 1983.
145. M. Mittelbach, B. Pokits, and A. Silberholz. Production and fuel properties of fatty acid methyl esters from used frying oil, In: *Liquid Fuels from Renewable Resources*, Nashville, Tennessee: American Society of Agricultural Engineers, 1992.
146. P. Felizardo, et al. Production of biodiesel from waste frying oils, *Waste Management* **26**, 487–494, 2006.
147. S. Zheng, et al. Acid-catalyzed production of biodiesel from waste frying oil, *Biomass and Bioenergy* **30**, 267–272, 2006.
148. M. I. Al-Widyan and A. O. Al-Shyoukh. Experimental evaluation of the transesterification of waste palm oil into biodiesel, *Bioresource Technology* **85**, 253–256, 2002.
149. A. A. Anggraini. *Wiederverwertung von gebrauchten Speiseölen/-fetten im energetisch-technischen Bereich-ein Verfahren und dessen Bewertung*, In: *Fachgebiet Agrartechnik*, 1999, Universität Gesamthochschule Kassel, Witzenhausen, Germany, p. 193.

150. Y. Wang, et al. Comparison of two different processes to synthesize biodiesel by waste cooking oil, *Journal of Molecular Catalysis A: Chemical* **252**, 107–112, 2006.
151. M. Canakci and J. Van Gerpen. Biodiesel production from oils and fats with high free fatty acids, *Transactions of the ASAE* **44**(6), 1429–1436, 2001.
152. Y. Watanabe, et al. Enzymatic conversion of waste edible oil to biodiesel fuel in a fixed-bed bioreactor, *JAOCS* **78**(7), 703–707, 2001.
153. R. Brenneis, B. Baeck, and B. Kley. Alcoholysis of waste fats with 2-ethyl-1-hexanol using *Candida antarctica* lipase A in large-scale tests, *European Journal of Lipid Science Technology* **106**, 809–814, 2004.
154. Z. F. Chen, H. Wu, and M. H. Zong. Transesterification of waste oil with high acid value to biodiesel catalyzed by immobilized lipase, *Chinese Journal of Catalysis* **27**(2), 146–150, 2006.
155. M. I. Al-Widyan, G. Tashtoush, and M. Abu-Qudais. Utilization of ethyl ester of waste vegetable oils as fuel in diesel engines, *Fuel Processing Technology* **76**, 91–103, 2002.
156. M. Mittelbach and H. Junek. *Verfahren zur Herstellung eines Fettsäureestergemisches aus Abfallfetten bzw. Ölen und Verwendung dieses Gemisches als Kraft-bzw. Brennstoff,* 1988, Austrian Patent 388 743 B, Austria.
157. M. Canakci and J. Van Gerpen. Comparison of engine performance and emissions for petroleum diesel fuel, yellow grease biodiesel, and soybean oil biodiesel, *Transactions of the ASAE* **46**(4), 937–944, 2003.
158. M. Çetinkaya, et al. Engine and winter road test performances of used cooking oil originated biodiesel, *Energy Conversion and Management* **46**, 1279–1291, 2005.
159. M. P. Dorado, et al. The effect of waste vegetable oil blend with diesel fuel on engine performance, *Transactions of the ASAE* **45**(3), 519–523, 2002.
160. B. Kegl and A. Hribernik. Experimental analysis of injection characteristics using biodiesel fuel, *Energy & Fuels* **20**(5), 2239–2248, 2006.
161. S. Kato, et al. Evaluation of ozone treated fish waste oil as a fuel for transportation, *Journal of Chemical Engineering in Japan* **37**(7), 863–870, 2004.
162. M. A. Dubé, A. Y. Tremblay, and J. Liu. Biodiesel production using a membrane reactor, *Bioresource Technology* **98**, 639–647, 2007.
163. B. Siedel, et al. Safety evaluation for a biodiesel process using prion-contaminated animal fat as a source, *Environmental Science & Pollution Research* **13**(2), 125–130, 2006.
164. F. R. Ma, L. D. Clements, and M. A. Hanna. Biodiesel fuel from animal fat. Ancillary studies on transesterification of beef tallow, *Industrial & Engineering Chemistry Research* **37**(9), 3768–3771, 1998.
165. M. S. Graboski and R. L. McCormick. Combustion of fat and vegetable oil derived fuels in diesel engines, *Progress in Energy and Combustion Science* **24**, 125–164, 1998.
166. M. Fangrui, L. D. Clements, and M. A. Hanna. The effect of mixing on transesterification of beef tallow, *Bioresource Technology* **69**, 289–293, 1998.
167. M. Gil, I. Jachmanian, and M. A. Grompone. Physicochemical properties of esters derived from beef tallow with different alcohols for its possible use as biodiesel, *Ingenieria Quimica* **24**, 23–32, 2003.
168. R. G. Nelson and M. D. Schrock. Energetic and economic feasibility associated with the production, processing, and conversion of beef tallow to a substitute diesel fuel, *Biomass and Bioenergy* **30**, 584–591, 2006.
169. S. Schober, I. Seidl, and M. Mittelbach. Ester content evaluation in biodiesel from animal fats and lauric oils, *European Journal of Lipid Science Technology* **108**, 309–314, 2006.
170. F. Ma, L. D. Clements, and M. A. Hanna. The effects of catalyst, free fatty acids, and water on transesterification of beef tallow, *Transactions of the ASAE* **41**(5), 1261–1264, 1998.
171. G. M. Tashtoush, M. I. Al-Widyan, and M. M. Al-Jarrah. Experimental study on evaluation and optimization of conversion of waste animal fat into biodiesel, *Energy Conversion and Management* **45**, 2697–2711, 2004.
172. B. A. Nebel and M. Mittelbach. Biodiesel from extracted fat out of meat and bone meal, *European Journal of Lipid Science Technology* **108**, 398–403, 2006.

173. K. T. Lee, T. A. Foglia, and K. S. Chang. Production of alkyl ester as biodiesel from fractionated lard and restaurant grease, *JAOCS* **79**(2), 191–195, 2002.
174. I. K. Mbaraka, K. J. McGuire, and B. H. Shanks. Acidic mesoporous silica for the catalytic conversion of fatty acids in beef tallow, *Industrial & Engineering Chemistry Research* **45**(9), 3022–3028, 2006.
175. V. T. Wyatt, et al. Fuel properties and nitrogen oxide emission levels of biodiesel produced from animal fats, *JAOCS* **82**(8), 585–591, 2005.
176. M. Foma and T. Abdala. Kernel oils of seven plant species of Zaire, *JAOCS* **62**(5), 910–911, 1985.
177. D. K. Abbiw. *Useful plants of Ghana*, London: Intermediate Technology Publication & Royal Botanic Gardens, Kew, 1990.
178. M. Lisa and M. Holcapek. Analysis of natural mixtures of triacylglycerols using HPLC/MS technique, *Chemické Listy* **99**, 195–199, 2005.
179. P. H. Mensier. *Dictionaire de huiles vegetales*, Paris: Editions Paul Lechevalier, 1949.
180. D. K. Sharma. Pharmacological properties of flavonoids including flavonolignans— Integration of petrocrops with drug development from plants, *Journal of Scientific & Industrial Research* **65**(6), 477–484, 2006.
181. J. W. Purseglove. *Tropical Crops: Monocotyledons*, Harlow, UK: Longman, 1985.
182. E. K. Marfo, O. L. Oke, and O. A. Afolabi. Nutritional evaluation of papaw (*Carica papaya*) and flamboyant (*Denolix regia*) seed oils, *Nutrition Reports International* **37**(2), 303–310, 1988.
183. K. C. Sen. Bulletin No. 25, Indian Council of Agricultural Research, 1938.
184. K. H. Gautam and N. N. Devoe. Ecological and anthropogenic niches of sal (*Shorea robusta* Gaertn. f.) forest and prospects for multiple-product forest management— a review, *Forestry* **79**(1), 81–101, 2006.
185. S. K. Behera and M. K. Misra. Aboveground tree biomass in a recovering tropical sal (*Shorea robusta* Gaertn. f.) forest of Eastern Ghats, India, *Biomass and Bioenergy* **30**, 509–521, 2006.
186. H. Brucher. *Useful Plants of Neotropical Origin and Their Wild Relatives*, Berlin Heidelberg: Springer-Verlag, p. 126, 1989.

Fuel and Physical Properties of Biodiesel Components

Gerhard Knothe

5.1 Introduction

Biodiesel is an alternative diesel fuel (DF) derived from vegetable oils or animal fats [1, 2]. Transesterification of an oil or fat with a monohydric alcohol, in most cases methanol, yields the corresponding mono-alkyl esters, which are defined as biodiesel. The successful introduction and commercialization of biodiesel in many countries around the world has been accompanied by the development of standards to ensure high product quality and user confidence. Some biodiesel standards are ASTM D6751 (ASTM stands for American Society for Testing and Materials) and the European standard EN 14214, which was developed from previously existing standards in individual European countries.

The suitability of any material as fuel, including biodiesel, is influenced by the nature of its major as well as minor components arising from production or other sources. The nature of these components ultimately determines the fuel and physical properties. Some of the properties included in standards can be traced to the structure of the fatty esters in the biodiesel. Since biodiesel consists of fatty acid esters, not only the structure of the fatty acids but also that of the ester moiety can influence the fuel properties of biodiesel. The transesterification reaction of an oil or fat leads to a biodiesel fuel corresponding in its fatty acid profile with that of the parent oil or fat. Therefore, biodiesel is largely a mixture of fatty esters with each ester component contributing to the properties of the fuel.

Properties of biodiesel that are determined by the structure of its component fatty esters and the nature of its minor components include ignition quality, cold flow, oxidative stability, viscosity, and lubricity. This chapter discusses the influence of the structure of fatty esters on these properties. Not all of these properties have been included in biodiesel standards, although all of them are essential to proper functioning of the fuel in a diesel engine.

Generally, as the least expensive alcohol, methanol has been used to produce biodiesel. Biodiesel, in most cases, can therefore be termed the fatty acid methyl esters (FAME) of a vegetable oil or animal fat. However, as mentioned above, both the fatty acid chain and the alcohol functionality contribute to the overall properties of a fatty ester. It is worthwhile to consider the properties imparted by other alcohols yielding fatty acid alkyl esters (FAAE) that could be used for producing biodiesel. Therefore, both structural moieties will be discussed in this chapter. Table 5.1 lists fuel properties of neat alkyl esters of fatty acids. Besides the fuel properties discussed here, the heat of combustion (HG) of some fatty compounds [3] is included in Table 5. 1, for the sake of underscoring the suitability of fatty esters as fuel with regard to this property.

TABLE 5.1   Properties of Fatty Acids and Esters[a]

Trivial (systematic) name; acronym[b]	MP[c] (°C)	BP[c,d] (°C)	Cetane no.	Viscosity[e]	HG[f], (kcal/mol)
Caprylic (Octanoic); 8:0	16.5	239.3			
Methyl ester		193	33.6 (98.6)[g]	0.99[j]; 1.19[k]	1313
Ethyl ester	−43.1	208.5		1.37 (25°)[j]	1465
Capric (Decanoic); 10:0	31.5	270	47.6 (98.0)[g]		1453.07 (25°)
Methyl ester		224	47.2 (98.1)[g]	1.40[j]; 1.72[k]	1625
Ethyl ester	−20	243–5	51.2 (99.4)[g]	1.99 (25°)[j]	1780
Lauric (Dodecanoic); 12:0	44	131[1]			1763.25 (25°)
Methyl ester	5	266[766]	61.4 (99.1)[g]	1.95[j]; 2.43[k]	1940
Ethyl ester	−1.8fr	163[25]		2.88[j]	2098
Myristic (Tetradecanoic); 14:0	58	250.5[100]			2073.91 (25°)
Methyl ester	18.5	295[751]	66.2 (96.5)[g]	2.69[j]	2254
Ethyl ester	12.3	295	66.9 (99.3)[g]		2406
Palmitic (Hexadecanoic); 16:0	63	350			2384.76 (25°)
Methyl ester	30.5	415–8[747]	74.5 (93.6)[g]; 85.9[i]	3.60[j]; 4.38[k]	2550
Ethyl ester	19.3/24	191[10]	93.1[i]		2717
Propyl ester	20.4	190[12]	85.0[i]		
Isopropyl ester	13–4	160[2]	82.6[i]		
Butyl ester	16.9		91.9[i]		
2-Butyl ester			84.8[i]		
Isobutyl ester	22.5, 28.9	199[5]	83.6[i]		
Stearic (Octadecanoic); 18:0	71	360[d]	61.7[h]		2696.12 (25°)

TABLE 5.1   Properties of Fatty Acids and Esters[a] *(Continued)*

Trivial (systematic) name; acronym[b]	MP[c] (°C)	BP[c,d] (°C)	Cetane no.	Viscosity[e]	HG[f], (kcal/mol)
Methyl ester	39	$442-3^{747}$	86.9 $(92.1)^g$; 101^i	4.74^j	2859
Ethyl ester	31–33.4	199^{10}	76.8^h; 97.7^i		3012
Propyl ester			69.9^h; 90.9^i		
Isopropyl ester			96.5^i		
Butyl ester	27.5	343	80.1^h; 92.5^i		
2-Butyl ester			97.5^i		
Isobutyl ester			99.3^i		
Palmitoleic (9(Z)-Hexadecanoic); 16:1					
Methyl ester			51.0^i		2521
Oleic (9(Z)-Octadecanoic); 18:1	16	286^{100}	46.1^h		2657.4 (25°)
Methyl ester	−20	218.5^{20}	55^h; 59.3^i	3.73^j; 4.51^k	2828
Ethyl ester		$216-7^{151}$	53.9^h; 67.8^i	5.50 $(25°)^j$	
Propyl ester			55.7^h; 58.8^i		
Isopropyl ester			86.6^i		
Butyl ester			59.8^h; 61.6^i		
2-Butyl ester			71.9^i		
Isobutyl ester			59.6^i		
Linoleic (9Z,12Z-Octadecadienoic); 18:2	−5	$229-30^{16}$	31.4^h		
Methyl ester	−35	215^{20}	42.2^h; 38.2^i	3.05^j; 3.65^k	2794
Ethyl ester		$270-5^{180}$	37.1^h; 39.6^i		
Propyl ester			40.6^h; 44.0^i		
Butyl ester			41.6^h; 53.5^i		
Linolenic (9Z,12Z,15Z-Octadecatrienoic); 18:3	−11		$230-2^{17}$	20.4^h	
Methyl ester					
Ethyl ester	−57/−52	$109^{0.018}$	20.6^g; 22.7^i	2.65^j; 3.14^k	2750
Propyl ester		$174^{2.5}$	26.7^h		
Butyl ester			26.8^h		
Ricinoleic (12-Hydroxy-9Z-octadecenoic); 18:1, 12-OH	5.5	245^{10}			
Methyl ester		$225-7^{15}$		15.44^k	
Erucic (13Z-Docosenoic); 22:1	33–4	265^{15}			
Methyl ester		$221-2^5$		5.91^j	3454
Ethyl ester		$229-30^5$			

[a] Adapted from Ref. [4].

[b] The numbers denote the number of carbons and double bonds. For example, in oleic acid, 18:1 stands for 18 carbons and 1 double bond.

[c] Melting point and boiling point data are from Refs. [5] and [6].

[d] Superscripts denote pressure (mm Hg) at which the boiling point was determined.

[e] Viscosity values determined at 40°C, unless indicated otherwise.

[f] HG values are from Refs. [3] and [5].

[g] Number in parentheses indicates purity (%) of the material used for CN determination as given in Ref. [7].

[h] Ref. [8].

[i] Ref. [9].

[j] Dynamic viscosity (mPa · s = cP), Ref. [10].

[k] Kinematic viscosity (mm²/s = cSt), Ref. [11].

5.2   Cetane Number and Exhaust Emissions

The cetane number (CN), which is related to the ignition properties, is a prime indicator of fuel quality in the realm of diesel engines. It is conceptually similar to the octane number used for gasoline. Generally, a compound that has a high octane number tends to have a low CN and vice versa. The CN of a DF is related to the ignition delay (ID) time, i.e., the time between injection of the fuel into the cylinder and onset of ignition. The shorter the ID time, the higher the CN, and vice versa.

Standards have been established worldwide for CN determination, e.g., ASTM D613 in the United States, and internationally the International Organization for Standardization (ISO) standard ISO 5165. A long straight-chain hydrocarbon, hexadecane ($C_{16}H_{34}$; trivial name cetane, giving the cetane scale its name) is the high-quality standard on the cetane scale with an assigned CN of 100. A highly branched isomer of hexadecane, 2,2,4,4,6,8,8-heptamethylnonane (HMN), a compound with poor ignition quality, is the low-quality standard with an assigned CN of 15. The two reference compounds on the cetane scale show that CN decreases with decreasing chain length and increasing branching. Aromatic compounds that are present in significant amounts in petrodiesel have low CNs but their CNs increase with increasing size of n-alkyl side chains [12, 13]. The cetane scale is arbitrary, and compounds with CN > 100 or CN < 15 have been identified. The American standard for petrodiesel (ASTM D975) prescribes a minimum CN of 40, while the standards for biodiesel prescribe a minimum of 47 (ASTM D6751) or 51 (European standard EN 14214). Due to the high CNs of many fatty compounds, which can exceed the cetane scale, the lipid combustion quality number for these compounds has been suggested [14].

The use of biodiesel reduces most regulated exhaust emissions from a diesel engine. The species reduced include carbon monoxide, hydrocarbons, and particulate matter (PM). Nitrogen oxide (NO_x) emissions are slightly increased, however. When blending biodiesel with petrodiesel, the effect of biodiesel is approximately linear to the blend level. A report summarizing exhaust emissions tests with biodiesel is available [15], and other summaries are given in Refs. [16, 17].

The structure of the fatty esters in biodiesel affects the levels of exhaust emissions. When using a 1991-model, 6-cylinder, 345-bhp (257-kW), direct-injection, turbocharged, and intercooled diesel engine, NO_x exhaust emission increased with increasing number of double bonds and decreasing chain length for saturated chains [18]. Although often a trade-off is observed between NO_x and PM exhaust emissions, no trade-off has been observed in this work when varying the chain length [18]. The CN and density were correlated with emission levels [18]. However, emissions are likely affected by the technology level of the engine. When conducting tests on a 2003-model, 6-cylinder, 14–L, direct-injection, turbocharged, intercooled diesel engine with exhaust gas recirculation (EGR), no chain length effect has been observed for NO_x exhaust emissions, although the level of saturation still played a significant role [19]. PM exhaust emissions were reduced to levels close to the US 2007 regulations required for ultra-low-sulfur petrodiesel fuel. Also, PM levels were lower than those for neat hydrocarbons which would be enriched in "clean" petrodiesel fuel [19]. In both studies [18, 19], NO_x emissions of the saturated esters were slightly below those of the reference petrodiesel fuel.

For petrodiesel fuel, higher CNs have been correlated with reduced NO_x exhaust emissions [20]. This correlation has led to efforts to improve the CN of biodiesel fuels by using additives known as cetane improvers [8]. Despite the inherent relatively high CNs of fatty compounds, NO_x exhaust emissions usually increase slightly when operating a diesel engine on biodiesel, as mentioned above. The relationship between the CN and engine emissions is complicated by many factors, including the technology level of the engine. Older, lower-injection pressure engines are generally very sensitive to CN, with increased CN causing significant reductions in NO_x emissions, due to shorter ID times and the resulting lower average combustion temperatures. More modern engines that are equipped with injection systems that control the rate of injection are not very sensitive to CN [21–23].

Historically, the first CN tests were carried out on palm oil ethyl esters [24, 25], which have a high CN, a result confirmed by later studies on many other vegetable oil-based DFs and individual fatty compounds. The influence of the compound structure on CNs of fatty compounds has been discussed in more recent literature [26], with the predictions made in that paper being confirmed by practical cetane tests [7–9, 13]. CNs of neat fatty compounds are given in Table 5.1. In summary, the results show that CNs decrease with increasing unsaturation and increase with increasing chain length, i.e., uninterrupted CH_2 moieties. However, branched esters derived from alcohols such as isopropanol have CNs competitive with methyl or other straight-chain alkyl esters [9, 27]. Thus, one long, straight chain suffices to impart a high CN, even if the other moiety is branched. Branched esters are of interest because they exhibit improved low-temperature properties.

Recently, cetane studies on fatty compounds have been conducted using the Ignition Quality Tester™ (IQT) [9]. The IQT is a further, automated development of a constant volume combustion apparatus (CVCA) [28, 29]. The CVCA was originally developed for determining CNs more rapidly with greater experimental ease, better reproducibility, reduced use of fuel, and therefore less cost than the ASTM D613 method utilizing a cetane engine. The IQT method, which is the basis of ASTM D6890, was shown to be reproducible and the results competitive with those derived from ASTM D613. Some results from the IQT are included in Table 5.1. For the IQT, ID and CN are related by the following equation [9]:

$$CN_{IQT} = 83.99 \, (ID - 1.512)^{-0.658} + 3.547 \qquad (5.1)$$

In the recently approved method ASTM D6890, which is based on this technology, only ID times of 3.6–5.5 ms [corresponding to 55.3–40.5 DCN (derived CN)] are covered as the precision may be affected outside that range. However, for fatty compounds, the results obtained by using the IQT are comparable to those obtained by other methods [9]. Generally, the results of cetane testing for compounds with lower CNs, such as more unsaturated fatty compounds, show better agreement over various related literature references than the results for compounds with higher CNs, because of the nonlinear relationship [see Eq. (5.1)] between the ID time and the CN, which was observed previously [30]. Thus, small changes at shorter ID times result in greater changes in CN than at longer ID times. This would indicate a leveling-off effect on emissions such as NO_x, as discussed above, once a certain ID time with corresponding CN has been reached as the formation of certain species depend on the ID time. However, for newer engines, this aspect must be modified as discussed above.

5.3   Cold-Flow Properties

One of the major problems associated with the use of biodiesel is poor low-temperature flow properties, documented by relatively high cloud points (CPs) and pour points (PPs) [1, 2]. The CP, which usually occurs at a higher temperature than the PP, is the temperature at which a fatty material becomes cloudy due to the formation of crystals and solidification of saturates. Solids and crystals rapidly grow and agglomerate, clogging fuel lines and filters and causing major operability problems. With decreasing temperature, more solids form and the material approaches the PP, the lowest temperature at which the material will still flow. Saturated fatty compounds have significantly higher melting points than unsaturated fatty compounds (Table 5.1), and in a mixture, they crystallize at higher temperatures than the unsaturates. Thus, biodiesel fuels derived from fats or oils with significant amounts of saturated fatty compounds will display higher CPs and PPs.

Besides the CP (ASTM D2500) and PP (ASTM D97) tests, two test methods for the low-temperature flow properties of petrodiesel exist, namely, the low-temperature flow test (LTFT) (used in North America; e.g., ASTM D4539) and cold filter plugging point (CFPP) (used outside North America; e.g., the European standard EN 116). These methods have also been used to evaluate biodiesel and its blends with No. 1 and

2 petrodiesel. Low-temperature filterability tests were stated to be necessary because of better correlation with operability tests than the CP or PP test [31]. However, for fuel formulations containing at least 10 vol% methyl esters, both LTFT and CFPP are linear functions of the CP [32]. Additional statistical analysis have shown a strong 1:1 correlation between LTFT and CP [32].

Several approaches to low-temperature problems of esters have been investigated, including blending with petrodiesel, winterization, additives, branched-chain esters, and bulky substituents in the chain. The latter approach may be considered a variation of the additive approach, as the corresponding compounds have been investigated in biodiesel at additive levels. Blending of esters with petrodiesel will not be discussed here.

Numerous, usually polymeric, additives were synthesized and reported mainly in the patent literature to have the effect of lowering the PP or sometimes even the CP. A brief overview of such additives has been presented [33]. Similarly, the use of fatty compound-derived materials with bulky moieties in the chain [34] at additive levels has been investigated. The idea associated with these materials is that the bulky moieties in these additives would destroy the harmony of the crystallizing solids. The effect of some additives appears to be limited because they more strongly affect the PP than the CP or they have only a slight influence on the CP. The CP, however, is more important than the PP for improving low-temperature flow properties [35].

The use of branched esters such as isopropyl, isobutyl, and 2-butyl esters instead of methyl esters [36, 37] is another approach for improving the low-temperature properties of biodiesel. Branched esters have lower melting points in the neat form (Table 5.1). These esters showed a lower T_{CO} (crystallization onset temperature), as determined by differential scanning calorimetry (DSC) for the isopropyl esters of soybean oil (SBO) by 7–11°C and for the 2-butyl esters of SBO by 12–14°C [36]. The CPs and PPs were also lowered by the branched-chain esters. For example, the CP of isopropyl soyate was given as −9°C [7] and that of 2-butyl soyate as −12°C [36]. In comparison, the CP of methyl soyate is around 0°C [32]. However, in terms of economics, only isopropyl esters appear attractive as branched-chain esters, although even they are more expensive than methyl esters. Branching in the ester chain does not have any negative effect on the CN of these compounds, as discussed above.

Winterization [35, 38, 39] is based on the lower melting points of unsaturated fatty compounds than saturated compounds (Table 5.1). This method removes by filtration the solids formed during the cooling of the vegetable oil esters, leaving a mixture with a higher content of unsaturated fatty esters and thus with lower CP and PP. This procedure can be repeated to further reduce the CPs and PPs. Saturated fatty compounds, which have higher CNs (Table 5.1) than unsaturated fatty compounds, are among the major compounds removed by winterization. Thus the CN of biodiesel decreases during winterization. Loss of material was reduced when winterization was carried out in presence of cold-flow improvers or solvents such as hexane and isopropanol [39].

In other work [40], tertiary fatty amines and amides have been reported to be effective in enhancing the ignition quality of biodiesel without negatively affecting the low-temperature properties. Also, saturated fatty alcohols of chain lengths $>C_{12}$ increased the PP substantially. Ethyl laurate weakly decreased the PP.

5.4 Oxidative Stability

Oxidative stability of biodiesel has been the subject of considerable research [41–62]. This issue affects biodiesel primarily during extended storage. The influence of parameters such as presence of air, heat, traces of metal, antioxidants, and peroxides as well as nature of the storage container was investigated in the aforementioned studies. Generally, factors such as the presence of air, elevated temperatures, or the presence of metals facilitate oxidation. Studies performed with the automated oil stability index (OSI) method have confirmed the catalyzing effect of metals on oxidation; however, the influence of the compound structure of the fatty esters, especially unsaturation as discussed below, was even greater [52]. Numerous other methods, including not only wet-chemical ones such as the acid value and peroxide value, but also pressurized differential scanning calorimetry, nuclear magnetic resonance (NMR), and so forth, have been applied in oxidation studies of biodiesel.

Two simple methods for assessing the quality of stored biodiesel are the acid value and viscosity since both increase continuously with increasing fuel degradation, i.e., deteriorating fuel quality. The peroxide value is less suitable because it reaches a maximum and then can decrease again due to the formation of secondary oxidation products [48].

Autoxidation occurs due to the presence of double bonds in the chains of many fatty compounds. Autoxidation of unsaturated fatty compounds proceeds with different rates, depending on the number and position of double bonds [63]. Especially the positions allylic to double bonds are susceptible to oxidation. The bis-allylic positions in common polyunsaturated fatty acids, such as linoleic acid (double bonds at .C-9 and .C-12, giving one bis-allylic position at C-11) and linolenic acid (double bonds at .C-9, .C-12, and C-15, giving two bis-allylic positions at C-11 and C-14), are even more prone to autoxidation than the allylic positions. The relative rates of oxidation given in the literature [63] are 1 for oleates (methyl, ethyl esters), 41 for linoleates, and 98 for linolenates. This is essential because most biodiesel fuels contain significant amounts of esters of oleic, linoleic, or linolenic acids, which influence the oxidative stability of the fuels. The species formed during the oxidation process cause the fuel to eventually deteriorate.

A European standard (EN 14112; Rancimat method) for oxidative stability has been included in the American and European biodiesel standards (ASTM D6751 and EN 14214). Both biodiesel standards call for determining oxidative stability at 110°C; however, EN 14214 prescribes a minimum induction time of 6 h by the Rancimat method while ASTM D6751 prescribes 3 h. The Rancimat method is nearly identical to the OSI method, which is an AOCS (American Oil Chemists' Society) method.

Besides preventing exposure of the fatty material to air, adding antioxidants is a common method to address the issue of oxidative stability. Common antioxidants are synthetic materials such as *tert*-butylhydroquinone (TBHQ), butylated hydroxytoluene (BHT), butylated hydroxyanisole (BHA), and propyl gallate (PG) as well as natural materials such as tocopherols. Antioxidants delay oxidation but do not prevent it, as oxidation will commence once the antioxidant in a material has been consumed.

5.4.1 Iodine value

The iodine value (IV) has been included in the European biodiesel standards to purportedly address the issue of oxidative stability and the propensity of the oil or fat to polymerize and form engine deposits. The IV is a measure of the total unsaturation of a fatty material measured in grams of iodine per 100 g of sample when formally adding iodine to the double bonds. An IV of 120 has been specified in EN 14214 and 130 in EN 14213, which would largely exclude vegetable oils such as soybean and sunflower oils as biodiesel feedstock. Thus the IV has not been included in biodiesel standards in the United States and Australia, and is limited to 140 in the South African standard (which would permit sunflower and soybean oils); the provisional Brazilian standard requires that it only be noted.

The IV of a vegetable oil or animal fat is almost identical to that of the corresponding methyl esters; however, the IV of alkyl esters decreases with higher alcohols used in their production since the IV is molecular weight dependent. For example, the IV of methyl, ethyl, propyl, and butyl linoleate is 172.4, 164.5, 157.4, and 150.8, respectively [64].

The use of the IV of a mixture for such purposes does not take into consideration that an infinite number of fatty acid profiles can yield the same IV and that different fatty acid structures can give the same IV, although the propensity for oxidation can differ significantly [64]. Other, new structure indices termed allylic and bis-allylic position equivalents (APE and BAPE), which are based on the number of such positions in a fatty acid chain and are independent of molecular weight, are likely more suitable than the IV [64]. The BAPE index distinguishes mixtures having nearly identical IV correctly by their OSI times. Note that the BAPE index is the decisive index compared to the APE because it relates to the more reactive bis-allylic positions. Engine performance tests with a mixture of vegetable oils of different IVs did not yield results that would have justified a low IV [65, 66]. No relationship between the IV and oxidative stability has been observed in another investigation on biodiesel with a wide range of IV [52].

5.5 Viscosity

Viscosity affects the atomization of a fuel upon injection into the combustion chamber and, thereby, ultimately the formation of engine deposits. The higher the viscosity, the greater the tendency of the fuel to cause such problems. The viscosity of a transesterified oil, i.e., biodiesel, is about an order of magnitude lower than that of the parent oil [1, 2]. High viscosity is the major fuel property why neat vegetable oils have been largely abandoned as alternative DF. Kinematic viscosity has been included in most biodiesel standards. It can be determined by standards such as ASTM D445 or ISO 3104. The difference in viscosity between the parent oil and the alkyl ester derivatives can be used in monitoring biodiesel production [67]. The effect on viscosity of blending biodiesel and petrodiesel has also been investigated [68], and an equation has been derived, which allows calculating the viscosity of such blends.

The prediction of viscosity of fatty materials has received considerable attention in the literature. Viscosity values of biodiesel/mixtures of fatty esters have been predicted from the viscosities of the individual components by a logarithmic equation for dynamic viscosity [10]. Viscosity increases with chain length (number of carbon atoms) and with increasing degree of saturation. This holds also for the alcohol moiety as the viscosity of ethyl esters is slightly higher than that of methyl esters [11]. Factors such as double bond configuration influence viscosity (cis double bond configuration giving a lower viscosity than the trans configuration), while the double bond position affects viscosity less [11]. Thus, a feedstock such as used frying oils, which is more saturated and contains some amounts of trans fatty acid chains, has a higher viscosity than its parent oil. Branching in the ester moiety, however, has little or no influence on viscosity, again showing that this is a technically promising approach for improving low-temperature properties without significantly affecting other fuel properties. Values for dynamic viscosity and kinematic viscosity of neat fatty acid alkyl esters are included in Table 5.1.

5.6   Lubricity

With the advent of low-sulfur petroleum-based DFs, the issue of DF lubricity is becoming increasingly important. Desulfurization of petrodiesel reduces or eliminates the inherent lubricity of this fuel, which is essential for proper functioning of vital engine components such as fuel pumps and injectors. Several studies [10, 11, 67–82] on the lubricity of biodiesel or fatty compounds have shown a beneficial effect of these materials on the lubricity of petrodiesel, particularly low-sulfur petrodiesel fuel. Adding biodiesel at low levels (1–2%) restores the lubricity to low-sulfur petroleum-derived DFs. However, the lubricity-enhancing effect of biodiesel at low blend levels is mainly caused by minor components of biodiesel such as free fatty acids and monoacylglycerols [83], which have free COOH and OH groups. Other studies [84, 85] also point out the beneficial effect of minor components on biodiesel lubricity, but these studies do not fully agree on the responsible species [83–85]. Thus, biodiesel is required at 1–2% levels in low-lubricity petrodiesel, in order for the minor components to be effective lubricity enhancers [83]. At higher blend levels, such as 5%, the esters are sufficiently effective without the presence of minor components.

While the length of a fatty acid chain does not significantly affect lubricity, unsaturation enhances lubricity slightly; thus an ester such as methyl linoleate or methyl linolenate improves lubricity more than methyl stearate [80, 83]. In accordance with the above observation on the effect of free OH groups on lubricity, castor oil displayed better lubricity than other vegetable oil esters [75, 80, 81]. Ethyl esters have improved lubricity compared to methyl esters [75].

Standards for testing DF lubricity use the scuffing load ball-on-cylinder lubricity evaluator (SLBOCLE) (ASTM D6078) or the high-frequency reciprocating rig (HFRR) (ASTM D6079; ISO 12156). Lubricity has not been included in biodiesel standards despite the definite advantage of biodiesel over petrodiesel with respect to this fuel property. However, the HFRR method has been included in the petrodiesel standards ASTM D975 and EN 590.

5.7   Outlook

The fuel properties of biodiesel are strongly influenced by the properties of the individual fatty esters as well as those of some minor components. Both moieties, the fatty acid and alcohol, have considerable influence on fuel properties such as CN, with relation to combustion and exhaust emissions, cold flow, oxidative stability, viscosity, and lubricity. It therefore appears reasonable to enrich (a) certain fatty ester(s) with desirable properties in the fuel, in order to improve the properties of the whole fuel. For example, from the presently available data, it appears that isopropyl esters have better fuel properties than methyl esters. The major disadvantage is the higher price of isopropanol in comparison to methanol, besides modifications needed for the transesterification reaction. Similar observations likely hold for the fatty acid moiety. It may be possible in the future to improve the properties of biodiesel by means of genetic engineering of the parent oils, which could eventually lead to a fuel enriched with (a) certain fatty acid(s), possibly oleic acid, that exhibits a combination of improved fuel properties.

References

1. G. Knothe, J. Krahl, and J. Van Gerpen (Eds.). *The Biodiesel Handbook*, Champaign, IL: AOCS Press, 2005.
2. M. Mittelbach and C. Remschmidt. *Biodiesel—The Comprehensive Handbook*, Graz, Austria: Martin Mittelbach, 2004.
3. B. Freedman and M. O. Bagby. Heats of combustion of fatty esters and triglycerides, *Journal of the American Oil Chemists' Society* 66, 1601–1605 (1989).
4. G. Knothe. Dependence of biodiesel fuel properties on the structure of fatty acid alkyl esters, *Fuel Processing Technology* 86, 1059–1070, 2005.
5. R. C. Weast (Ed.). *Handbook of Chemistry and Physics*, 66th ed., Boca Raton, FL: CRC Press, p. D272, 1986.
6. F. D. Gunstone, J. L. Harwood, and F. B. Padley. *The Lipid Handbook*, London: Chapman & Hall, 1994.
7. W. E. Klopfenstein. Effect of molecular weights of fatty acid esters on cetane numbers as diesel fuels, *Journal of the American Oil Chemists' Society* **62**, 1029–1031, 1985.
8. G. Knothe, M. O. Bagby, and T. W. Ryan III. Cetane Numbers of Fatty Compounds: Influence of Compound Structure and of Various Potential Cetane Improvers, SAE Technical Paper Series 971681, In: *State of Alternative Fuel Technologies*, SAE Publication SP-1274, Warrendale, PA, pp. 127–132, 1997.
9. G. Knothe, A. C. Matheaus, and T. W. Ryan III. Cetane numbers of branched and straight-chain fatty esters determined in an ignition quality tester, *Fuel* **82**, 971–975, 2003.
10. C. A. W. Allen, K. C. Watts, R. G. Ackman, and M. J. Pegg. Predicting the viscosity of biodiesel fuels from their fatty acid ester composition, *Fuel* **78**, 1319–1326, 1999.
11. G. Knothe and K. R. Steidley. Kinematic viscosity of biodiesel fuel components. Influence of compound structure and comparison to petrodiesel fuel components, *Fuel* **84**, 1059–1065, 2005.
12. A. D. Puckett and B. H. Caudle. Ignition qualities of hydrocarbons in the diesel-fuel boiling range, *U.S. Bureau of Mines Information Circular*, No. 7474, 14 pp., 1948.
13. P. Q. E. Clothier, B. D. Aguda, A. Moise, and H. Pritchard. How do diesel-fuel ignition improvers work? *Chemical Society Reviews* **22**, 101–108, 1993.
14. B. Freedman, M. O. Bagby, T. J. Callahan, and T. W. Ryan III. Cetane Numbers of Fatty Esters, Fatty Alcohols and Triglycerides Determined in a Constant Volume Combustion Bomb, SAE Technical Paper Series 900343, 1990.
15. U.S. Environmental Protection Agency. A Comprehensive Analysis of Biodiesel Impacts on Exhaust Emissions, Draft Technical Report EPA420-P-02-00, October 2002; see also www.epa.gov/oms/models/analysis.p02001.pdf (accessed January 2007).
16. R. L. McCormick and T. L. Alleman. Effect of biodiesel fuel on pollutant emissions from diesel engines, In: *The Biodiesel Handbook*, Knothe, G., Van Gerpen, J., and Krahl, J., (Eds.), Champaign, IL: AOCS Press, pp. 165–174, 2005.
17. J. Krahl, A. Munack, O. Schröder, H. Stein, and J. Bünger. Influence of biodiesel and different petrodiesel fuels on exhaust emissions and health effects, In: *The Biodiesel Handbook*, Knothe, G., Van Gerpen, J., and Krahl, J. (Eds.), Champaign, IL: AOCS Press, pp. 175–182, 2005.

18. R. L. McCormick, M. S. Graboski, T. L. Alleman, and A. M. Herring. Impact of biodiesel source material and chemical structure on emissions of criteria pollutants from a heavy-duty engine, *Environmental Science and Technology* **35**, 1742–1747, 2001.
19. G. Knothe, C. A. Sharp, and T. W. Ryan III. Exhaust emissions of biodiesel, petrodiesel, neat methyl esters, and alkanes in a new technology engine, *Energy & Fuels* **20**, 403–408, 2006.
20. N. Ladommatos, M. Parsi, and A. Knowles. The effect of fuel cetane improver on diesel pollutant emissions, *Fuel* **75**, 8–14, 1996.
21. R. L. Mason, A. C. Matheaus, T. W. Ryan III, R. A. Sobotowski, J. C. Wall, C. H. Hobbs et al. EPA HDEWG Program—Statistical Analysis, SAE Technical Paper Series 2001-01-1859, In: *Diesel and Gasoline Performance and Additives*, SAE Publication SP-1551, Warrendale, PA, 2001.
22. A. C. Matheaus, G. D. Neely, T. W. Ryan III, R. A. Sobotowski, J. C. Wall, C. H. Hobbs, G. W. Passavant, and T. J. Bond. EPA HDEWG Program—Engine Test Results, SAE Technical Paper Series 2001-01-1858, In: *Diesel and Gasoline Performance and Additives*, SAE Publication SP-1551, Warrendale, PA, 2001.
23. R. A. Sobotowski, J. C. Wall, C. H. Hobbs, A. C. Matheaus, R. L. Mason, T. W. Ryan III, et al. EPA HDEWG Program—Test Fuel Development, SAE Technical Paper Series 2001-01-1857, In: *Diesel and Gasoline Performance and Additives*, SAE Publication SP-1551, Warrendale, PA, 2001.
24. M. van den Abeele. Palm oil as raw material for the production of a heavy motor fuel (L'huile de palme. Matière première pour la préparation d'un carburant lourd utilisable dans les moteurs à combustion interne.), *Bulletin Agricole du Congo Belge* **33**, 3–90, 1942.
25. G. Knothe. Historical perspectives on vegetable oil-based diesel fuels, *INFORM* **12**, 1103–1107, 2001.
26. K. J. Harrington. Chemical and physical properties of vegetable oil esters and their effect on diesel fuel performance, *Biomass* **9**, 1–17, 1986.
27. Y. Zhang and J. H. Van Gerpen. Combustion Analysis of Esters of Soybean Oil in a Diesel Engine, SAE Technical Paper Series 960765, In: *Performance of Alternative Fuels for SI and CI Engines*, SAE Publication SP-1160, Warrendale, PA, pp. 1–15, 1996.
28. T. W. Ryan III and B. Stapper. Diesel Fuel Ignition Quality as Determined in a Constant Volume Combustion Bomb, SAE Technical Paper Series 870586, 1987.
29. A. A. Aradi and T. W. Ryan III. Cetane Effect on Diesel Ignition Delay Times Measured in a Constant Volume Combustion Apparatus, SAE Technical Paper Series 952352, In: *Emission Processes and Control Technologies in Diesel Engines*, SAE Publication SP-1119, Warrendale, PA, pp. 43–56, 1995.
30. L. N. Allard, G. D. Webster, N. J. Hole, T. W. Ryan III, D. Ott, and C. W. Fairbridge. Diesel Fuel Ignition Quality as Determined in the Ignition Quality Tester (IQT), SAE Technical Paper Series 961182, 1996.
31. K. Owen and T. Coley. *Automotive Fuels Reference Book*, Second ed., Warrendale, PA: SAE, 1995.
32. R. O. Dunn and M. O. Bagby. Low-temperature properties of triglyceride-based diesel fuels: Transesterified methyl esters and petroleum middle distillate/ester blends, *Journal of the American Oil Chemists' Society* **72**, 895–904, 1995.
33. G. Knothe, R. O. Dunn, and M. O. Bagby. Biodiesel: The use of vegetable oils and their derivatives as alternative diesel fuels, ACS Symposium Series 666, In: *Fuels and Chemicals from Biomass*, American Chemical Society, Washington, DC, pp. 172–208, 1997.
34. G. Knothe, R. O. Dunn, M. W. Shockley, and M. O. Bagby. Synthesis and characterization of some long-chain diesters with branched or bulky moieties, *Journal of the American Oil Chemists' Society* **77**, 865–871, 2000.
35. R. O. Dunn, M. W. Shockley, and M. O. Bagby. Improving the low-temperature properties of alternative diesel fuels: Vegetable-oil derived methyl esters, *Journal of the American Oil Chemists' Society* **73**, 1719–1728, 1996.
36. I. Lee, L. A. Johnson, and E. G. Hammond. Use of branched-chain esters to reduce the crystallization temperature of biodiesel, *Journal of the American Oil Chemists' Society* **72**, 1155–1160, 1995.

37. T. A. Foglia, L. A. Nelson, R. O. Dunn, and W. Marmer. Low-temperature properties of alkyl esters of tallow and grease, *Journal of the American Oil Chemists' Society* **74**, 951–955, 1997.
38. I. Lee, L. A. Johnson, and E. G. Hammond. Reducing the crystallization temperature of biodiesel by winterizing methyl soyate, *Journal of the American Oil Chemists' Society* **73**, 631, 1996.
39. R. O. Dunn, M. W. Shockley, and M. O. Bagby. Winterized Methyl Esters from Soybean Oil: An Alternative Diesel Fuel with Improved Low-Temperature Flow Properties, SAE Technical Paper Series 971682, In: *State of Alternative Fuel Technologies*, SAE Publication SP-1274, Warrendale, PA, p. 133, 1997.
40. S. Stournas, E. Lois, and A. Serdari. Effects of fatty acid derivatives on the ignition quality and cold flow of diesel fuel, *Journal of the American Oil Chemists' Society* **72**, 433–437, 1995.
41. L. M. Du Plessis. Plant oils as diesel fuel extenders: Stability tests and specifications on different grades of sunflower seed and soyabean oils, *CHEMSA* **8**, 150–154, 1982.
42. L. M. Du Plessis, J. B. M. de Villiers, and W. H. van der Walt. Stability studies on methyl and ethyl fatty acid esters of sunflowerseed oil, *Journal of the American Oil Chemists' Society* **62**, 748–752, 1985.
43. P. Bondioli, A. Gasparoli, A. Lanzani, E. Fedeli, S. Veronese, and M. Sala. Storage stability of biodiesel, *Journal of the American Oil Chemists' Society* **72**, 699–702, 1995.
44. P. Bondioli and L. Folegatti. Evaluating the oxidation stability of biodiesel. An experimental contribution, *Rivista Italiana delle Sostanze Grasse* **73**, 349–353, 1996.
45. N. M. Simkovsky and A. Ecker. Influence of light and contents of tocopherol on the oxidative stability of fatty acid methyl esters (Einfluß von Licht und Tocopherolgehalt auf die Oxidationsstabilität von Fettsäuremethylestern.), *Fett/Lipid* **100**, 534–538, 1998.
46. J. C. Thompson, C. L. Peterson, D. L. Reece, and S. M. Beck. Two-year storage study with methyl and ethyl esters of rapeseed, *Transactions of the American Society of Agricultural Engineers* **41**, 931–939, 1998.
47. N. M. Simkovsky and A. Ecker. Effect of antioxidants on the oxidative stability of rapeseed oil methyl esters, *Erdoel, Erdgas, Kohle* **115**, 317–318, 1999.
48. M. Canakci, A. Monyem, and J. Van Gerpen. Accelerated oxidation processes in biodiesel, *Transactions of the American Society of Agricultural Engineers* **42**, 1565–1572, 1999.
49. R. O. Dunn. Analysis of oxidative stability of methyl soyate by pressurized-differential scanning calorimetry, *Transactions of the American Society of Agricultural Engineers* **43**, 1203–1208, 2000.
50. M. Mittelbach and S. Gangl. Long storage stability of biodiesel made from rapeseed and used frying oil, *Journal of the American Oil Chemists' Society* **78**, 573–577, 2001.
51. R. O. Dunn. Effect of oxidation under accelerated conditions on fuel properties of methyl soyate (biodiesel), *Journal of the American Oil Chemists' Society* **79**, 915–920, 2002.
52. G. Knothe and R. O. Dunn. Dependence of oil stability index of fatty compounds on their structure and concentration and presence of metals, *Journal of the American Oil Chemsits' Society* **80**, 1021–1026, 2003.
53. S. Schober and M. Mittelbach. The impact of antioxidants on biodiesel oxidation stability, *European Journal of Lipid Science and Technology* **106**, 382–389, 2004.
54. P. Bondioli, A. Gasparoli, L. della Bella, S. Tagliabue, F. Lacoste, and L. Lagardere. The prediction of biodiesel storage stability. Proposal for a quick test, *European Journal of Lipid Science and Technology* **106**, 822–830, 2004.
55. T. Dittmar, B. Ondruschka, J. Haupt, and M. Lauterbach. Improvement of the oxidative stability of fatty acid methyl esters with antioxidants—Limitations of the Rancimat test (Verbesserung der Oxidationsstabilität von Fettsäuremethylestern mit Antioxidantien—Grenzen des Rancimat-Tests.), *Chemie Ingenieur Technik* **76**, 1167–1170, 2004.
56. O. Falk and R. Meyer-Pitroff. The effect of fatty acid composition on biodiesel oxidative stability, *European Journal of Lipid Science and Technology* **106**, 837–843, 2004.
57. R. O. Dunn. Oxidative stability of soybean oil fatty acid methyl esters by oil stability index (OSI), *Journal of the American Oil Chemists' Society* **82**, 381–387, 2005.

58. R. O. Dunn. Effect of antioxidants on the oxidative stability of methyl soyate (biodiesel), *Fuel Processing Technology* **86**, 1071–1085, 2005.
59. J. Polavka, J. Paligová, J. Cvengroš, and P. Šimon. Oxidation stability of methyl esters studied by differential thermal analysis and Rancimat, *Journal of the American Oil Chemists' Society* **82**, 519–524 (2005).
60. R. O. Dunn. Oxidative stability of biodiesel by dynamic mode pressurized-differential scanning calorimetry (P-DSC), *Transactions of the American Society of Agricultural and Biological Engineers* **49**, 1633–1641, 2006.
61. H. L. Fang and R. L. McCormick. Spectroscopic Study of Biodiesel Degradation Pathways, SAE Technical Paper Series 2006-01-3300, 2006.
62. G. Knothe. Analysis of oxidized biodiesel by ¹H-NMR and effect of contact area with air, *European Journal of Lipid Science and Technology* **108**, 493–500, 2006.
63. E. N. Frankel. *Lipid Oxidation*, Brigdwater, England: The Oily Press, 2005.
64. G. Knothe. Structure indices in FA chemistry. How relevant is the iodine value? *Journal of the American Oil Chemists' Society* **79**, 847–854, 2002.
65. H. Prankl, M. Wörgetter, and J. Rathbauer. Technical performance of vegetable oil methyl esters with a high iodine number, In: *Biomass, Proceedings of the 4th Biomass Conference of the Americas*, 1999, Oakland, California, pp. 805–810.
66. H. Prankl and M. Wörgetter. Influence of the iodine number of biodiesel to the engine performance, In: *Liquid Fuels and Industrial Products from Renewable Resources, Proceedings of the 3rd Liquid Fuel Conference*, Cundiff, J. S., et al. (Eds.), St Joseph, MI: ASAE, pp. 191–196, 1996.
67. P. De Filippis, C. Giavarini, M. Scarsella, and M. Sorrentino. Transesterification processes for vegetable oils: A simple control method of methyl ester content, *Journal of the American Oil Chemists' Society* **71**, 1399–1404, 1995.
68. M. E. Tat and J. H. Van Gerpen. The kinematic viscosity of biodiesel and its blends with diesel fuel, *Journal of the American Oil Chemists' Society* **76**, 1511–1513, 1999.
69. J. A. Waynick. Evaluation of the stability, lubricity, and cold flow properties of biodiesel fuel, In: *Proceedings of the 6th International Conference on Stability and Handling of Liquid Fuels*, 1997, Vancouver, BC, Canada.
70. J. H. Van Gerpen, S. Soylu, and M. E. Tat. Evaluation of the lubricity of soybean oil-based additives in diesel fuel, In: *Proceedings of 1999 ASAE/CSAE-SCGR Annual International Meeting*, Paper No. 996134, Toronto, Canada, 1999.
71. D. Karonis, G. Anostopoulos, E. Lois, S. Stournas, F. Zannikos, and A. Serdari. Assessment of the Lubricity of Greek Road Diesel and the Effect of the Addition of Specific Types of Biodiesel, SAE Technical Paper Series 1999-01-1471, 1999.
72. L.G. Schumacher and J. Van Gerpen. Engine Oil Analysis of Diesel Engines Fueled with 0, 1, 2 and 100 Percent Biodiesel, In: ASAE Meeting Presentation 006010, *ASAE International Meeting*, 2000, Midwest Express Center, Milwaukee, WI.
73. G. Anastopoulos, E. Lois, A. Serdari, F. Zanikos, S. Stornas, and S. Kalligeros. Lubrication properties of low-sulfur diesel fuels in the presence of specific types of fatty acid derivatives, *Energy & Fuels* **15**, 106–112, 2001.
74. G. Anastopoulos, E. Lois, D. Karonis, F. Zanikos, and S. Kalligeros. A preliminary evaluation of esters of monocarboxylic fatty acid on the lubrication properties of diesel fuel, *Industrial & Engineering Chemistry Research* **40**, 452–456, 2001.
75. D. C. Drown, K. Harper, and E. Frame. Screening vegetable oil alcohol esters as fuel lubricity enhancers, *Journal of the American Oil Chemists' Society* **78**, 579–584, 2001.
76. C. Kajdas and M. Majzner. The Influence of Fatty Acids and Fatty Acids Mixtures on the Lubricity of Low-Sulfur Diesel Fuels, SAE Technical Paper Series 2001-01-1929, 2001.
77. L. G. Schumacher and B. T. Adams. Using biodiesel as a lubricity additive for petroleum diesel fuel, *ASAE Paper*, No. 02-6085, 2002.
78. A. K. Agarwal, J. Bijwe, and L. M. Das. Effect of biodiesel utilization of wear of vital parts in compression ignition engine, *Transactions of the American Society of Mechanical Engineers (Journal of Engineering for Gas Turbines and Power)* **125**, 604–611, 2003.
79. A. K. Agarwal, J. Bijwe, and L. M. Das. Wear assessment in a biodiesel fueled compression ignition engine, *Transactions of the American Society of Mechanical Engineers (Journal of Engineering for Gas Turbines and Power)* **125**, 820–826, 2003.

80. D. P. Geller and J. W. Goodrum. Effects of specific fatty acid methyl esters on diesel fuel lubricity, *Fuel* **83**, 2351–2356, 2004.
81. J. W. Goodrum and D. P. Geller. Influence of fatty acid methyl esters from hydroxylated vegetable oils on diesel fuel lubricity, *Bioresource Technology* **96**(7), 851–855, 2005.
82. A. K. Bhatnagar, S. Kaul, V. K. Chhibber, and A. K. Gupta. HFRR studies on methyl esters of nonedible vegetable oils, *Energy & Fuels* **20**(3), 1341–1344, 2006.
83. G. Knothe and K. R. Steidley. Lubricity of components of biodiesel and petrodiesel. The origin of biodiesel lubricity, *Energy & Fuels* **19**, 1192–1200, 2005.
84. G. Hillion, X. Montagne, and P. Marchand. Methyl esters of plant oils used as additives or organic fuel. (Les esters méthyliques d'huiles végétales: additif ou biocarburant?), *Oleagineux, Corps Gras, Lipides* **6**, 435–438, 1999.
85. J. Hu, Z. Du, C. Liu, and E. Min. Study on the lubrication properties of biodiesel as fuel lubricity enhancers, *Fuel* **84**, 1601–1606, 2005.

Processing of Vegetable Oils as Biodiesel and Engine Performance

Ahindra Nag

6.1 Introduction

Processing of vegetable oils as biodiesel [1, 2] and its engine performance is very challenging. From an environmental point of view, diesel engines are a major source of air pollution. Exhaust gases from diesel engines contain oxides of nitrogen, carbon monoxide, organic compounds consisting of unburned or partially burned hydrocarbons and particulate matter (consisting primarily of soot).

Interest in clean burning fuels is growing worldwide, and reduction in exhaust emissions from diesel engines is of utmost importance. It is widely recognized that alternative diesel fuels produced from vegetable oils and animal fats can reduce exhaust emissions from compression ignition (CI) engines, without significantly affecting engine performance. But reducing pollutant emissions from diesel engines requires a detailed knowledge of the combustion process. However, the complex nature of the combustion process in an engine makes it difficult to understand the events occurring in the combustion chamber that determine the emission of exhaust gases.

Dr. Rudolf Diesel [3], the inventor of the CI engine, used peanut oil in one of his engines for a demonstration at the Paris exhibition in 1900. Then there was considerable interest in the use of vegetable oils as fuel in diesel engines.

Several studies have reported the effects of fuel and engine parameters on diesel exhaust emissions. Chowdhury [4] claims to have successfully used raw vegetable oils in diesel engines. He observed that no major changes were necessary in the engine, but the engine could not be run for more than 4 h. The performance and economic aspects of vegetable oil were also discussed.

Barve and Amurthe [5] cite an example of using groundnut oil as fuel in a diesel engine generator set (103 kW) of a local water pump house. They claimed that the power output and fuel consumption were very much comparable with certified diesel fuel. Weibe and Nowakowska [6] have reported the use of palm oil as a motor fuel. The performance was found satisfactory with higher fuel consumption. Fang [7] has reported that soybean and castor oil blended with diesel fuel burns adequately in a small diesel engine. Engelman et al. [8] has presented data on the performance of soybean diesel oil blends compared with diesel fuel. Results from a short-duration test showed that the use of blends was feasible in the diesel engine; but in fact, in the long-term, test problems associated with lubrication, sticking piston rings, and injector atomization patterns contributed to mechanical difficulties in the engine. Cruze et al. [9] have found that atomization of the fuel by the injector, in some cases, has caused delayed ignition characteristics and reduced efficiency of mechanical power production, compared to diesel fuel. Pryde [10] has stated that raw vegetable oil has had no great promise for engine tests and that modified oil esters were required for further engine tests. Bruwer [11] has reported that even without modification, nine diesel engines started and operated almost normally on sunflower oil and delivered power equal to that of diesel fuel. Brake thermal efficiency and maximum engine power were 3% lower, while the specific fuel consumption was 10% higher than that of diesel fuel. The bench test, however, showed that atomization of 100% sunflower oil was much poorer than diesel but could be improved by reducing the viscosity of oil. Energy wise, sunflower oil was favorable for running diesel engines for a shorter duration.

Baranescu et al. [12] have conducted tests on a turbocharged engine, using mixtures of sunflower oil in 25%, 50%, and 75% with diesel fuel. They have concluded that the use of sunflower oil blended with diesel brought modification in the fuel injection process that mainly included an increase in injection pressure and a longer ignition duration. These effects led to longer combustion duration. Cold-temperature operation was very critical due to high viscosity that caused fuel system problems such as starting failure, unacceptable emission levels, and injection pump failure. Engine shutdown for a long duration accelerated gum formation, where the fuel contacted the bare metal. This might further impair the engine or injection system.

Wagner et al. [13] have conducted tests on a number of diesel engines with different blends of winter rape and safflower oil with diesel fuel. The following specific conclusions were drawn from the results obtained:

- High viscosity and tendency to polymerize within the cylinder were major physical and chemical problems.

- Attempt to reduce the viscosity of the oil by preheating the fuel by increasing the temperature of the fuel at the injector to the required value was not successful.

- Short-term engine performance showed power output and fuel consumption equivalent to diesel fuel.

- Severe engine damage occurred within a very short duration when the test was conducted for maximum power with varying engine rpm (revolutions per minute).

- A blend of 70% winter rape with 30% diesel was successfully used for 50 h. No adverse effect was noted.

- A diesel injector pump when run for 154 h with safflower oil had no abnormal wear, gumming, or corrosion.

Borgelt et al. [14] have conducted tests on three diesel engines containing 25–75% and 50–50% soybean oil and diesel. The engines were operated under 50% load for 1000 horsepower (HP); the output ranged from 2.55 to 2.8 kW. Thermal efficiency ranged from 19.3 to 20%. Engine performances were not significantly different. Carbon deposit increased with increased percentage of soybean oil. Thus, Borgelt et al. concluded that use of 25% or less soybean oil caused negligible changes in engine performance.

Barsic and Humke [15] performed a study in which blends of unrefined peanut and sunflower with diesel fuel (50–50%) were used in a single-cylinder engine. The engine produced equivalent power or a minor increase (6%) with vegetable oils and blends, with a 20% increase in specific fuel consumption. Performance tests at equal energy showed that the power level remained constant or decreased slightly, thermal efficiency decreased slightly, and the exhaust temperature increased with an increase in the percentage of vegetable oil in the fuel. Exhaust emission at equal energy input showed slightly higher NO_x for vegetable oils and their blends. Unburned hydrocarbon emission was about 50% higher than pure diesel fuel because the injection system was not optimized for more viscous fuels. Ziejewski et al. [16] reported the results of an endurance test using a 25–75% blend of alkali-refined sunflower oil with diesel and 25–75% blend of safflower oil with diesel on a volume basis. The major problems experienced were premature injection, determination of nozzle performance, and heavier carbon deposits in the piston ring grooves.

There was no significant problem with engine operation when the blend of safflower oil was used. That investigation revealed that chemical differences between vegetable oil and diesel had a very important influence on long-term engine performance. Bhattacharya et al. [17] have reported that a blend of 50% rice bran oil with diesel could be a supplementary fuel for their 10-bhp CI engine. No significant difference in the brake thermal efficiency was reported.

Samson et al. [18] have reported the use of tallow and stillingia oil in 25–75% and 50–50% blends by mass with diesel. The fuel properties of the blends were found to be within the limits proposed for diesel. The heat of combustion appeared to decrease. Specific gravity and kinematic viscosity increased with the increase in concentration of oil. Dunn et al. [19] conducted the test on rubber seed oil blended with diesel in 25%, 50%, 75%, and 100% in an air-cooled engine with 4.9 kW at 3600 rpm. Higher specific fuel consumption and slightly higher thermal efficiency were observed. But, carbon deposits were heavier than those for pure diesel fuel.

Samga [20] conducted a test on a water-cooled single-cylinder diesel engine, using hone oil (ken seed oil). He concluded that hone oil gave acceptable performance, smooth running, and ease in starting without preheating. The exhaust temperature and specific fuel consumption were higher than those for diesel. The partial-load efficiency was lower, but full-load efficiency was better than with diesel fuel.

Auld et al. [21] evaluated the potential yield and fuel qualities of winter rape, safflower, and sunflower as sources of fuel for diesel engines. Vegetable oils contained 94–95% heat value of diesel fuel, but were 11.1–17.6 times more viscous and also 7–9% heavier than diesel fuel. Viscosities of vegetable oils were closely related to fatty acid chain length and number of unsaturated bonds. During short-term engine tests, all vegetable oils produced power comparable to that of diesel, and the thermal efficiency was 1.8–2.8% higher than that of diesel. Based on the results, they concluded that vegetable oil as fuel should be selected on identification of the crop species that produced the most optimum yield of fuel quality vegetable oils.

Ryan et al. [22] have tested four different types of vegetable oils (soybean, sunflower, cottonseed, and peanut) in at least three different stages of processing. All the oils were characterized according to their physical and chemical properties. The spray characteristics of oils were determined at different fuel temperatures, using a high-pressure, high-temperature injection bomb, and high-speed motion picture camera. The injection study pointed out that vegetable oil behaved differently from diesel fuel. Normally, as the viscosity decreased, the penetration rate decreased and the spray cone angle increased. Using vegetable oils, however, increased the penetration rate, and increasing the temperature of the oil from 45°C to 145°C reduced the cone angle and decreased the viscosity.

Engine test results, based on the specific energy, showed that degummed soybean performed as well as the base fuel, but performance of the deodorized sunflower was the worst of those tested with an energy consumption 10% higher than the base fuel. Vegetable oils had a much smaller premixed combustion stage, with the diffused stage of combustion being flatter for sunflower and soybean oil than for the diesel fuel. Engine inspection showed that heating of the oil reduced the carbon deposit problem. It was concluded that deposits and overall durability were related to viscosity differences and the chemical structures of the other oil as compared to diesel fuel.

Mathur and Das [23] have conducted tests on diesel engines, using blends of mahua and neem oil with diesel. Results showed that neem oil could be substituted for up to 35% with marginal reduction in efficiency and power output. Mahua oil with diesel had exhaust characteristics similar to those of diesel. Further, savings in the diesel fuel through the use of both these nonedible oils outweighed the demerits of a marginal drop in efficiency and a slight loss in power output.

Goering et al. [24] conducted tests on a diesel engine using a hybrid fuel formed by micro-emulsion of aqueous ethanol in soybean oil. The test data were compared with the data from a baseline test on diesel fuel. The nonionic emulsion produced the same power as diesel fuel, with 19% lower heating value. Brake specific fuel consumption (BSFC) was 16% higher, and the brake thermal efficiency was 6% higher, with diesel at full power. Diesel knock for the hybrid fuel was not worse than for diesel fuel; thus the low cetane number of the hybrid fuel was not reflected in engine performance. Hybrid fuels were less volatile than ethanol and thus safer. The effect of hybrid fuel on the engine durability was unknown.

6.2   Processing of Vegetable Oils to Biodiesel

Different techniques adopted for converting vegetable oils to biodiesel are (a) degumming of vegetable oils, (b) transesterification by acid or alkali, and (c) enzymatic transesterification.

6.2.1   Degumming of vegetable oils

Degumming is an economical chemical process involving acid treatment to improve the viscosity and cetane number up to a certain limit so that the blends of nonedible oils with diesel can be used satisfactorily in a diesel engine. It is a very simple process by which the gum of the vegetable oil is removed to decrease the viscosity of oil by using an appropriate acid that can be optimized for reduction in viscosity. The quantity of acid and the duration of the process are very important to obtain optimum results. Compared to transesterification, the process of degumming is simple, very easy, and less costly, and the reduction in viscosity of vegetable oil is very small.

Nag et al. [25] degummed karanja, putranjiva, and jatropha oils by phosphoric acid treatment. Before degumming the oils, the fuel properties of three oils have been measured and compared with diesel (Table 6.1). Acid concentrations of 1%, 2%, 3%, 4%, and 5% were used at 40°C with vigorous stirring. The stirring was continued for 10 min after adding the acid. After stirring, the mixtures were held for 1 week to complete the reactions and to settle the gum materials. Then the mixtures were filtered through a packed bed filled with charred sawdust. Viscosities of the filtrate were then measured.

Performance and emission measurement.   After studying the properties of the jatropha, karanja, and putranjiva oils, they were degummed. In this context, the Ricardo variable-compression engine (Ricardo & Co. Engineers Ltd., England, single cylinder, 3-in bore, 35/8 in stroke) was run with 10%, 20%, 30%, and 40% blends of degummed karanja, jatropha, and putranjiva oils with diesel at different loads (0–2.7 kW) and different timings (45°, 40°, 35°, and 30° bTDC [before top dead center]). To measure emissions, an automotive exhaust monitor (model PEA205) and smoke meter (model OMS103, Indus Scientific Pvt. Ltd., India) were used.

Degumming by acid treatment lowers the viscosity. Viscosities of karanja, jatropha, and putranjiva oils degummed at 40°C and at various acid concentrations are shown in Fig. 6.1. Karanja oil with 4% acid treatment had the lowest viscosity, whereas jatropha and putranjiva oils both had the lowest viscosities with 1% acid treatment.

Effect of timing.   By observing the performance data at various timings (45°, 40°, and 35° bTDC) in Fig. 6.2, it was concluded that at 45° bTDC timing, the nonedible karanja, jatropha, and putranjiva oils gave the highest yields, whereas at 40° bTDC timing, diesel gave the highest yield. That may have been due to the different ignition temperatures of the nonedible oils from diesel.

TABLE 6.1   Fuel Properties of Three Nonedible Oils and Diesel

Properties	Karanja	Jatropha	Putranjiva	Diesel
Viscosity in cSt (at 40°C)	43.67	35.38	37.62	5.032
Cetane number	29.9	33.7	31.3	46.3
Calorific value (kJ/kg)	36,258	38,833	39,582	42,707
Pour point (°C)	5	2	−3	−12
Specific gravity at 25°C	0.932	0.916	0.918	0.834
Flash point (°C)	215	280	48	78
Fire point (°C)	235	291	53	85
Carbon residue (%)	1.4	0.2	0.9	0.1

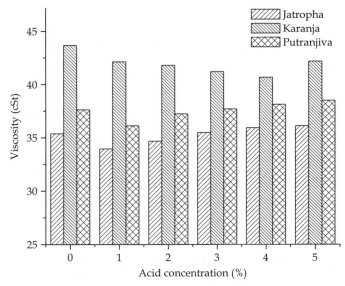

Figure 6.1   Viscosity versus acid concentration of jatropha, karanja, and putranjiva oils at 40°C.

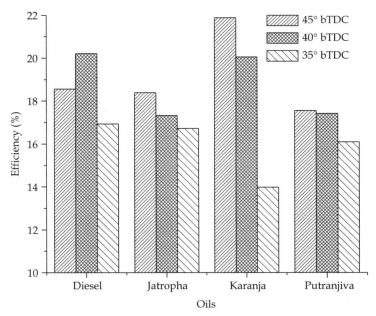

Figure 6.2   Brake thermal efficiency at various timings of diesel and 20% vegetable oil blends at 1-kW brake power, 1200 rpm, and 20 compression ratio.

Performance of various blends.   Performances of blends of degummed vegetable oil with diesel are shown in Figs. 6.3 and 6.4. The 20% blends of jatropha, karanja, and putranjiva oils with diesel gave quite satisfactory performance related to BSFC and brake thermal efficiency (η_{bt}). Beyond the 20% blends, the cetane numbers and viscosities of the blends were not so effective.

Comparison of the performance of blends.   As per Figs. 6.5 and 6.6, engine performance using jatropha and karanja oils was better than diesel but the use of putranjiva oil gave reverse results at all loads, although the results were more or less the same. Degummed karanja oil blends gave better performance, but at high loads, the performance of jatropha oil blends was better in comparison to the performance of karanja oil blends. The performance data showed that all three vegetable oils could be used as alternative fuels for diesel engines.

Effect of loads on emissions of vegetable oil blends and comparison.   As per Figs. 6.7 and 6.8, it is interesting to note that for the karanja, jatropha, and putranjiva oils, in every case, smoke and particulates decreased, which was very favorable in terms of their environmental impact on human beings. The rate of increase in smoke and particulate generation with the load of jatropha oil, in comparison to karanja and putranjiva

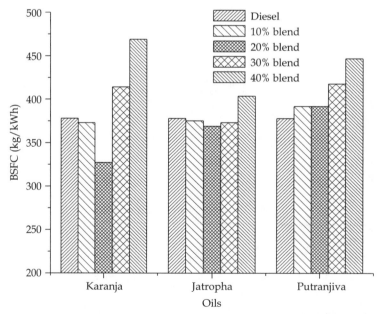

Figure 6.3 Brake specific fuel consumption versus vegetable oils and diesel blends at 1200 rpm, 45° bTDC, 20 compression ratio, and 1.4-kW brake power.

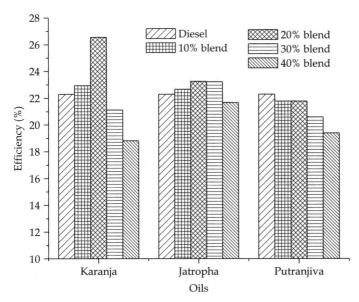

Figure 6.4   Brake thermal efficiency versus brake horsepower of vegetable oil and diesel blends at 1200 rpm, 45° bTDC, 20 compression ratio, and 1.4-kW brake power.

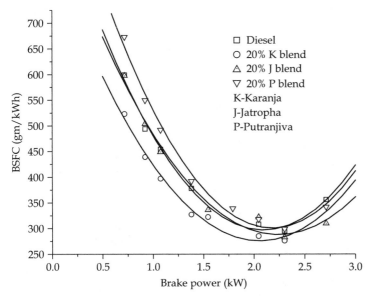

Figure 6.5   Brake specific fuel consumption versus brake power of diesel, 20% karanja oil, jatropha oil, and putranjiva oil blends at 1200 rpm, 45° bTDC, and 20 compression ratios.

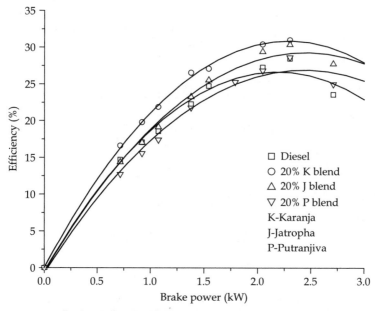

Figure 6.6 Brake thermal efficiency versus brake power of diesel, 20% karanja oil, 20% jatropha oil, and 20% putranjiva oil blends at 1200 rpm, 45° bTDC, and 20 compression ratio.

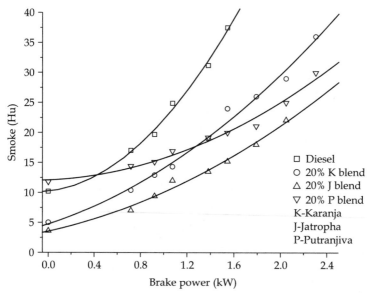

Figure 6.7 Smoke versus brake power of diesel, 20% karanja oil, 20% jatropha oil, and 20% putranjiva oil blends at 1200 rpm, 45° bTDC, and 20 compression ratio.

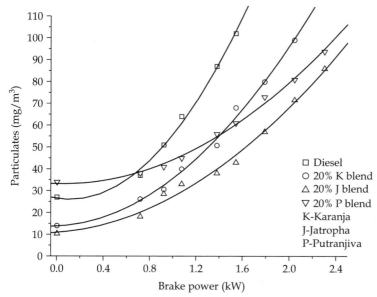

Figure 6.8 Particulates versus brake power of diesel, 20% karanja oil, 20% jatropha oil and 20% putranjiva oil blends at 1200 rpm, 45° bTDC, and 20 compression ratio.

oils, was very low. It is very interesting to observe that although the particulates and smoke for all the oils decreased, jatropha oil blends gave the highest reduction.

In Figs. 6.9 and 6.10, the CO, CO_2, NO_x, and HC (hydrocarbon) emissions for the three nonedible oils were less in comparison to diesel at high loads. However, at low loads, emissions from the nonedible oils are almost parallel to diesel. Because of the higher ignition temperature of nonedible oils than diesel, the better combustion of these oils gave less exhaust emissions.

Thus, degumming is an economic chemical process for a 20% blend of karanja, jatropha, and putranjiva oils with diesel to have very satisfactory results. The degumming method, therefore, offers a potential low-cost method with simple technology for producing an alternative fuel for CI engines. Out of the three nonedible oils, jatropha oil was the most promising to yield good performance and emissions at high loads in all respects. Comparing CO, CO_2, NO_x, HC, smoke, and particulate emissions from using the three nonedible oils, jatropha oil was very encouraging (see Fig. 6.11). Considering the above-mentioned points, it can be concluded that the diesel engine can be run very satisfactorily using a 20% blend of vegetable oil with diesel at 45° bTDC, 1200 rpm, and 20 compression ratios. Any diesel engine can be operated with a 20% blend

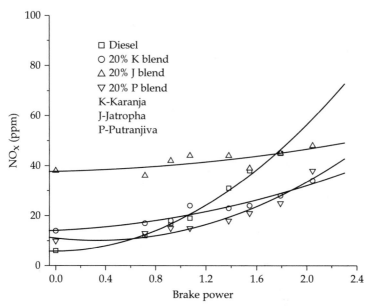

Figure 6.9 Nitrogen oxide versus brake power of diesel, 20% karanja oil, 20% jatropha oil, and 20% putranjiva oil blends at 1200 rpm, 45° bTDC, and 20 compression ratio.

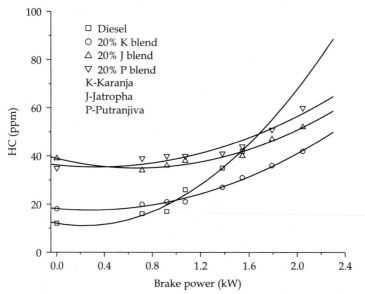

Figure 6.10 Unburnt hydrocarbon versus brake power of diesel, 20% karanja oil, 20% jatropha oil, and 20% putranjiva oil blends at 1200 rpm, 45° bTDC, and 20 compression ratio.

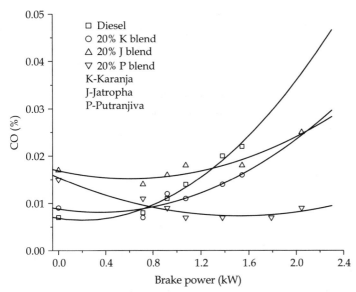

Figure 6.11 Carbon monoxide versus brake power of diesel, 20% karanja oil, 20% jatropha oil, and 20% putranjiva oil blends at 1200 rpm, 45° bTDC, and 20 compression ratio.

of degummed vegetable oils as a prime mover for agriculture purposes without any modification of the engine.

6.2.2 Transesterification of vegetable oils by acid or alkali

Goering et al. [24] have suggested that vegetable oils are too viscous for prolonged use in direct-injected diesel engines, which has led to poor fuel atomization and inefficient mixing with air, contributing to incomplete combustion. These chemical and physical properties caused vegetable oils to accumulate and remain as charred deposits when they contacted engine cylinder walls. The problem of charring and deposits of oils on the injector and cylinder wall can be overcome by better esterification of the oil to reduce the viscosity and remove glycerol.

Acid-catalyzed alcoholysis of triglycerides (TG) can be used to produce alkyl esters for a variety of traditional applications and for potentially large markets in the biodiesel fuel industry [26]. It can overcome some of the shortcomings of traditional base catalysis for producing alkyl esters. A significant disadvantage of base catalysts is their inability to esterify free fatty acids (FFA). These FFA are present at about 0.3 wt% in refined soybean oil and at significantly higher concentrations in waste greases, due to hydrolysis of the oil with water to produce FFA. The FFA react with soluble bases to form soaps through the saponification reaction mechanism. The soap forms emulsions and makes recovery of methyl esters (ME) difficult. Saponification consumes the base catalyst and reduces product yields. The use of alkaline catalysts requires that the oil reagent be dry and contain less than about 0.3 wt% FFA [27, 28].

Acid catalysts can handle large amounts of FFA and are commonly used to esterify FFA in fat or oil feedstock prior to base-catalyzed FFA alcoholysis to ME [29]. Though it solves FFA problems, it adds additional reaction and cleanup steps that increase batch times, catalyst cost, and waste generation.

Generally, acid-catalyzed methanolysis of TG is carried out at temperatures at or below that of methanol reflux (65°C). Using sulfuric acid catalysis under reflux conditions, Harrington and D'Arcy-Evans [30] first explored the feasibility of in situ transesterification, using homogenized whole sunflower seeds as a substrate. Using reflux conditions, a 560-fold molar excess of methanol and a 12-fold molar excess of sulfuric acid relative to the number of moles of triacylglycerol (TAG) were used. They observed ester production, with yields up to 20% greater than in the transesterification of preextracted oil, and suggested that this was an effect of the water content of the seeds, an increased extractability of some seed lipids under acidic conditions, and also the transesterification of seed-hull lipids.

Stern et al. [31] have developed a process to prepare ethyl esters for use as a diesel fuel substitute from various vegetable oils using hydrated ethyl alcohol and crude vegetable oil, with sulfuric acid as a catalyst. Ethyl ester of 98% purity with a very low acidity has been reported.

Schwab et al. [32] have compared acid and base catalysts and confirmed that, although base catalysts performed well at lower temperatures, acid catalysis requires higher temperatures. Liu [33] has compared the influence of acid and base catalysts on yield and purity of the product, and suggested that an acid catalyst is more effective for alcoholysis if the vegetable oil contains more than 1% FFA.

Goff et al. [34] have conducted acid-catalyzed alcoholysis of soybean oil using sulfuric, hydrochloric, formic, acetic, and nitric acids, which were evaluated at 0.1 and 1 wt% loadings at temperatures of 100°C and 120°C in sealed ampoules, and observed sulfuric acid was effective. Kinetic studies at 100°C with 0.5 wt% sulfuric acid catalyst and 9 times methanol stoichiometry provided more than 99 wt% conversion of TG in 8 h, and with less than 0.8 wt% FFA concentration in less than 4 h (see Fig. 6.12).

Base catalysts are generally preferred to acid catalysts because they lead to faster reactions [35]. Base catalysts generally used in transesterification reactions are NaOH, KOH, and their alkoxides. KOH is preferred to other bases because the end reaction mixture can be neutralized with phosphoric acid, which produces potassium phosphate, a well-known fertilizer [36].

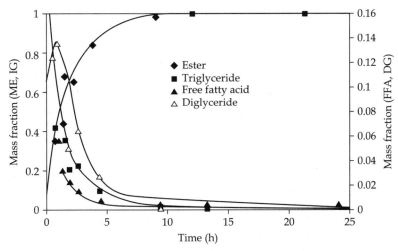

Figure 6.12 Kinetics of 0.5 wt% sulfuric acid catalyst at 100°C and 9:1 methanol-TG molar ratio. (*Used with permission from Goff et al. [34].*)

Darnoko et al. [37] explained transesterification of palm oil with methanol and KOH as a catalyst by the following three-step reaction sequence:

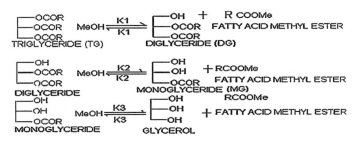

Knothe et al. [38] have reported optimal conditions of a 1 wt% KOH catalyst at 69°C and 7:1 alcohol—vegetable oil molar ratio gave 97.7% conversions in 18 min, when KOH was used with high-purity feedstocks.

Freedman et al. [39] have studied transesterification of sunflower oil and soybean oil with the reaction variables (a) molar ratio of alcohol to vegetable oil, (b) type of alcohol (methanol, ethanol, and *t*-butanol), (c) type of catalyst (acidic and alkali), and (d) reaction temperature (60°C, 45°C, and 32°C). They have suggested that esterification was 90–98% completed at the respective molar ratio of methanol to sunflower oil 4:1 and 6:1. All three alcohols produced high yields of esters. Alkaline catalysts were generally much more effective than acid catalysts. The reaction was performed successfully at both 45°C and 60°C in 4 h, with the production of 97% of ME.

Kruclen et al. [40] have presented a process for conversion of a high-melting point palm oil fraction into ethyl esters, which could be used as a diesel fuel substitute. The amount of catalyst used (KOH) was 0.1–1%, and the reaction was completed rapidly at 80°C with yields of 80–94%, depending on the concentration of catalysts. The specific gravity of ethyl ester varied from 0.847 to 0.864 with kinematic viscosity of 4.4–4.6 cSt at 40°C.

Gelbard et al. [41] have determined the yield of transesterification of rapeseed oil with methanol and base by ^{1}H-NMR (nuclear magnetic resonance) spectroscopy. The relevant signals chosen for integration are those of methoxy groups in ME at 3.7 ppm (parts per million) (singlet) and of the α-carbonyl methylene groups present in all fatty ester derivatives at 2.3 ppm. The latter appears as a triplet, so accurate measurements require good separation of this multiple at 2.1 ppm, which is related to allylic protons.

Chadha et al. [42] have studied base-catalyzed transesterification of monoglycerides from pongamia oil. They separated monoglyceride fractions (MG) by column chromatography and then characterized the fractions by ^{1}H-NMR spectroscopy in deuterated chloroform (CDCl$_3$) and tetramethylsilane (TMS) (see Fig. 6.13). They explain that 1- or 2-MG are positional isomers. Consequently, in 1-MG, the methylene protons at

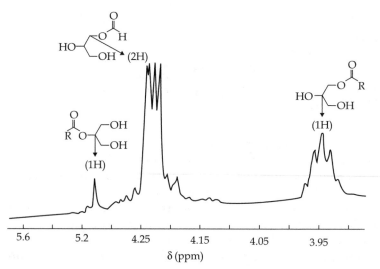

Figure 6.13 Characteristic ^{1}H-NMR signals of 1- and 2-MG. (*Used with permission from Chadha [42].*)

C-1 and C-3 are magnetically nonequivalent, due to four double doublets, which are observed in the spectra. But 2-MG, on the other hand, are symmetrical, and C-1 and C-3 methylene protons are magnetically equivalent and appear as a multiplate.

6.2.3 Enzymatic transesterification of vegetable oils

Enzymatic transesterification of TG by lipases (3.1.1.3) is a good alternative over a chemical process due to its eco-friendly, selective nature and low temperature requirement. Lipases break down the TAG into FFA and glycerol that exhibits maximum activity at the oil–water interface. Under low-water conditions, the hydrolysis reaction is reversible, i.e., the ester bond is synthesized rather than hydrolyzed. Scientists are interested in the development of lipase applications to the interesterification reactions of vegetable oils for production of biodiesel.

Nag has reported [43] celite-immobilized commercial *Candida rugosa* lipase and its isoenzyme lipase 4 efficiently catalyzed alcoholysis (dry ethanol) of various TG and soybean oil (see Fig. 6.14). This process has many advantages over chemical processes such as (a) low reaction temperature, (b) no restriction on organic solvents, (c) substrate specificity on enzymatic reactions, (d) efficient reactivity requiring only the mixing of the reactants, and (e) easy separation of the product.

Kaieda et al. have developed [44] a solvent-free method for methanolysis of soybean oil using *Rhizopus oryzae* lipase in the presence of 4–30 wt%

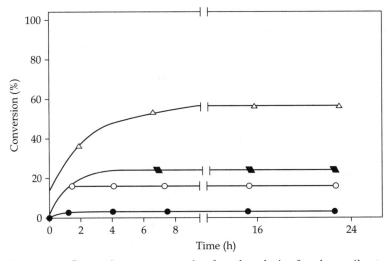

Figure 6.14 Conversion versus reaction for ethanolysis of soybean oil catalyzed by immobilized lipase 4 at 40°C and 250 rpm. Ethyl oleate (△); ethyl palmitate (◆); ethyl stearate (○); ethyl linoleate (•).

water in the starting materials. Oda et al. [45] have reported methanolysis of the same oil using whole-cell biocatalyst, where *R. oryzae* cells were immobilized within porous biomass support particles (BSP). Köse et al. [46] have reported the lipase-catalyzed synthesis of alkyl esters of fatty acids from refined cottonseed oil using primary and secondary alcohols in the presence of an immobilized enzyme from *C. antarctica*, commercially called Novozym-435 in a solvent-free medium. Under the same conditions, with short-chain primary and secondary alcohols, cottonseed oil was converted into its corresponding esters.

Alcoholysis of soybean oil with methanol and ethanol using several lipases has been investigated. The immobilized lipase from *Pseudomonas cepacia* was the most efficient for synthesis of alkyl esters, where 67 and 65 mol% of methyl and ethyl esters, respectively, were obtained by Noureddini et al. [47]. Shimada et al. [48] have reported transesterification of waste oil with stepwise addition of methanol using immobilized *C. antarctica* lipase, where they have successfully converted more than 90% of the oil to fatty acid ME. They have also implemented the same technique for ethanolysis of tuna oil.

Dossat et al. [49] have found that hexane was not a good solvent as the glycerol formed after the reaction was insoluble in *n*-hexane and adsorbed onto the enzyme, leading to a drastic decrease in enzymatic activity. Enzymatic transesterification of cottonseed oil has been studied using immobilized *C. antarctica* lipase as catalyst in *t*-butanol solvent by Royon et al. [50].

Sometimes, gums present in the oils used inhibit alcoholysis reactions due to interference in the interaction of the lipase molecule with substrates by the phospholipids present in the oil gum. Crude soybean oil cannot be transesterified by immobilized *C. antarctica* lipase. So, Watanabe et al. [51] have used degummed oil as a substrate for a transesterification reaction, in order to minimize this problem, and have effectively achieved conversion of 93.8% oil to biodiesel.

Methanol is insoluble in the oil, so it inhibits the lipases, thereby decreasing its catalytic activity toward the transesterification reaction. Du et al. [52] transesterified soybean oil using methyl acetate in the presence of Novozym-435 (see Fig. 6.15). Further, glycerol was also insoluble in the oil and adsorbed easily onto the surface of the immobilized lipase, leading to a negative effect on lipase activity. They have suggested that methyl acetate was a novel acceptor for biodiesel production and no glycerol was produced in that process, as shown below:

$$
\begin{array}{ccc}
CH_2\text{-OOC-}R_1 & & R_1\text{-COOCH}_3 \quad CH_2\text{-OOCCH}_3 \\
| & \xrightarrow{\text{Lipase}} & | \\
CH\text{-OOC-}R_2 + 3CH_3COOCH_3 & \rightleftharpoons & R_2\text{-COOCH}_3 + CH\text{-OOCCH}_3 \\
| & & | \\
CH_2\text{-OOC-}R_3 & & R_3\text{-COOCH}_3 \quad CH_2OOCCH_3
\end{array}
$$

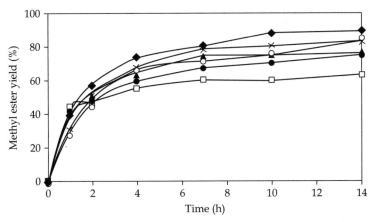

Figure 6.15 Effect of the substrate ratio of methyl acetate to oil as biodiesel production. Reaction conditions: 40°C; 150 ppm; 30% Novozym-435, based on oil–methyl acetate molar ratio 6:1 (□), 8:1 (•), 10:1 (▲), 12:1 (♦), and 24:1 (○). (*Used with permission from Du et al. [52].*)

They found 92% yield with a methyl acetate–oil molar ratio of 12:1, and methyl acetate showed no negative effect on enzyme activity.

The comparison of biodiesel production by acid, alkali, and enzyme is given in Table 6.2.

6.2.4   Engine performance with esters of vegetable oil

Hawkins et al. [53] have conducted combustion studies on methyl and ethyl esters of degummed sunflower oil, maize oil, cottonseed oil, peanut oil, soybean oil, and castor oil. Fuel properties of the esters were very similar to each other, except the esters of castor oil which were much more viscous. The heating values of ethyl esters were also considerably lower. Engine results indicated that the power output for esters varied from 44.4 to 45.5 kW, with diesel delivering 45.1 kW. The brake thermal efficiencies were also slightly higher than diesel. High esterification yields (around 90%) must be obtained to avoid choking of injector tips. Further, sticking of injector needles after a shutdown time of 48 h has been reported.

Fort and Blumberg [54] have tested a diesel engine with a mixture of cottonseed oil and ME of this oil. Results indicate that viscosity and density increased whereas the heating value and the cetane number decreased, when the percentage of the cottonseed oil was increased in the blend. The durability test with 50–50% cottonseed oil and ME was terminated after 183 h of running the engine, because the engine was noisy. After disassembly, the engine indicated severe wear scouring and

TABLE 6.2 Comparison of Biodiesel Production by Acid, Alkali, and Enzyme

Acid-catalyzed transesterification	Base-catalyzed transesterification	Enzymatic Transesterification
The glycerides and alcohol need not be anhydrous	The glycerides and alcohol must be substantially anhydrous	The alcohol needs to be anhydrous
Does not make a soap-like product	Soap formation taking place during the reaction	Does not make a soap-like product
Easy to water wash for product separation	Water washing is difficult due to soap emulsifier	Separation of the product is very easy; product is obtained only by filtration
Recommended for any free-fatty acid content of vegetable oil	Recommended for low free-fatty acid content of vegetable oil	Recommended for any free-fatty acid content of vegetable oil
Converts free fatty acid to ester	Converts free fatty acid to soap	Converts free fatty acid to ester
Product yield is high	Product yield is comparatively low	Product yield depends on different types of enzymes used; reaction is selective
Is slower than alkali-catalyzed transesterification	Is faster comparatively	Ezymatic reaction is slower than acid and alkali-catalyzed reaction
Percentage of conversion is low	Percentage of conversion is high	Percentage of conversion is high
Under low water content in reactants (oil or alcohol), transesterification reaction is not hindered	Water content in the reactant inhibits the reaction rate	Under low water conditions, the hydrolysis reaction is reversible, i.e., the ester bond is synthesized rather than hydrolyzed. Lipases break down the triacylglycerols into free fatty acids and glycerol that exhibit maximum activity at the oil–water interface
Need high temperature	Reacts even at room temperature	Conversion takes place at low temperature

heavy carbon deposits. But specific emissions and visible smoke characteristics of diesel fuel and esterified cottonseed oil were comparable.

Ziejewski and Kaufman [55] conducted a long-term test using a 25–75% blend of alkali-refined sunflower oil and diesel fuel in a diesel engine, and compared the results with that of a baseline test on diesel fuel. Engine power output over the tested speed range was slightly higher for this blend. At 2300 rpm, the difference was 25%. At 1800 rpm, the gain in power was 6%. The smoke level increased at a higher engine speed from 1 to 2.2 and decreased at a lower engine speed. Greater exhaust temperature was caused by a higher intake air temperature. The major problems experienced were:

1. Abnormal carbon buildup in the injection nozzle tips.

2. Injector needle sticking.

3. Secondary injection.

4. Carbon buildup in the intake port and exhaust-valve stems.

5. Carbon filling of the compression ring grooves.

6. Abnormal lacquer and varnish buildup.

Tahir [56] has determined the fuel properties of sunflower oil and its ME. The properties were favorable for diesel engine operation, but the problem of high viscosity (14 times higher than diesel at 37°C) of sunflower oil might cause blockage of fuel filters, higher valve-opening pressure, and poor atomization in the combustion chamber. Transesterification of sunflower oil to its ME has been suggested to reduce viscosity of the fuel. The viscosity of ME at 0°C was closer to that of No. 2 diesel fuel, but below 0°C, it was not possible because of the pour point of −4°C.

Pryor et al. [57] have conducted a short-term performance test on a small, test diesel engine using crude soybean oil, crude degummed soybean oil, and soybean ethyl ester. The engine developed about 3% more power output with crude legume soybean oil, but the development was insignificant with soybean ethyl ester. The fuel flow of soybean oil was 13–30% higher and for the ethyl ester it was 11–15% higher, depending upon the load on the engine. The exhaust temperature throughout the test was 2–5% higher for soybean oil and 2–3% lower for ethyl ester than the diesel fuel.

Clark et al. [58] have tested methyl and ethyl esters of soybean oil as a fuel in CI engine. Esters of soybean oil with commercial diesel fuel additives revealed fuel properties comparable to diesel fuel, with the exception of gum formation which manifested itself in problems with the plugging of fuel filters. Engine performance with esters differed little from the diesel fuel performance. Emissions of nitrous oxides for the esters were similar, or slightly higher than diesel fuel. Measurement of engine wear and the fuel injection test showed no abnormal characteristics for any of the fuels after 200 h of testing.

Laforgia et al. [59] has prepared biodiesel from degummed vegetable oil with 99.5% methanol and an alkaline catalyst (KOH). On engine performance, pure biodiesel and blends of biodiesel combined with 10% methanol had a remarkable reduction in smoke emissions. When the injection timing was advanced, better results were obtained.

Pischinger et al. [60] have conducted engine and vehicle tests with ME of soybean oil (MESO) 75–25% gas oil—MESO blend and 68–23–9% gas oil—MESO—ethanol blend. The fuel properties of the blend indicated a 6% lower volumetric calorific value of the ester, a drastic reduction in kinematics viscosity, and a greater ethane number than that of gas oil. The engine results indicated about 7% higher BSFC with a marginal difference in power and torque in comparison with gas oil. The smoke emission was much lower with ME.

Ali et al. [61] have observed that engine performance with diesel fuel—methyl soyate blends did not differ to a great extent up to a 70–30% (v/v) from that of diesel-fueled engine performance. There was a slight increase in NO_x emissions with increasing methyl soyate content in the blends at higher speeds but at lower speeds there was a quadratic trend with diesel fuel content.

Carbon monoxide emissions were very similar for blends up to 70–30% (v/v) diesel fuel—methyl soyate blends at any speed. Visible smoke decreased with increasing speed and methyl soyate content. More smoke was produced with neat diesel fuel at full load.

6.3 Engine Performance with Esters of Tallow and Frying Oil

The estimated amount of good quality and nutritive-value oils and fats used for frying around the world is around 20 million metric tons (MT). In frying, the hot oil serves as a heat exchange medium by which heat is transferred to the material being fried. As a result of frying, the oil darkens from the formation of polar materials such as minor phenolic components; elevated FFA; high total polar materials; compounds having high foaming property, low smoke point, low iodine value, and increased viscosity; and color compounds.

Sims [62] reported has conversion of tallow, a by-product of the meat industry, into esters. The fuel properties of methyl, ethyl, and butyl esters of tallow were similar to diesel fuel, particularly ME, which were remarkably similar except for the higher liquidification temperature of tallow esters. Short-term engine performance tests with methyl, ethyl, and butyl esters gave comparable results as diesel fuel, but at higher BSFC. Blends with diesel in 50–50% proportion by volume gave intermediate results between esters and neat diesel fuel.

Richardson et al. [63] have tested an engine with ME of tallow. Preliminary engine tests indicated that the use of 10% and 20% blends (volume basis) performed similar to diesel fuel. However, lubricant quality aspects were not studied and an endurance test was not conducted. The ignition quality of the blend was significantly better than that of diesel. Overall, it was concluded that tallow ME on 10% (volume basis) can be successfully used as diesel fuel where large amounts of tallow are produced and temperatures below 10°C are not encountered. The fuel consumption of ME of used frying oil has been measured by Mittelbach and Tritthard [64]. The ester fuel showed slightly lower hydrocarbon and carbon monoxide emissions but increased oxides of nitrogen, compared with that of diesel fuel. The particulate emissions, however, were significantly lower for used frying oil. But, they suggest long-term engine testing to prove the quality of this fuel.

The results discussed contribute to a better understanding of the structure—physical property relationships in different fatty acid esters from different vegetable oils which give the desired biodiesel quality and optimal performance of engines.

References

1. D. L. Klass. *Biomass for Renewable Energy, Fuels and Chemicals,* New York: Academic Press, 1998
2. V. V. Kafarov. *Wasteless Chemical Processing,* Moscow, Russia: Mir Publishers, 1985.
3. W. R. Nitseke and C. M. Wilson. *Rudolf Diesel: Pioneer of the Age of Power,* Norman, UK: University of Oklahoma Press, 1965.
4. D. H. Chowdhury. Indian vegetable fuel oils for diesel engines, *Gas and Oil Power,* **37**(5), 80–85, 1942.
5. R. V. Barve and P. V. Amurthe. Ground nut oil for diesel engine, *Current Science* **2**(10), 403, 1942.
6. R. Weibe and J. Nowakowska. The technical literature of agricultural motor fuels, USDA Bibliographical Bulletin No. 10, Superintendent of Documents, US Government Printing Office, Washington, DC, 1949.
7. K. S. Fang. Vegetable Oils and Diesel Fuel for China, Unpublished M.S. Thesis, University of Nebraska, Lincoln, NE.
8. H. W. Engelman, D. A. Guenther, and T. W. Silvis. Vegetable Oil as a Diesel Fuel, ASME, New York, ASME Paper No. 78-DG-P-19, November 5, 1978.
9. Z. M. Cruz, A. S. Ogunlowo, W. J. Chancellor, and J. R. Goss. Vegetable oils as fuels for diesel engines, *Resources and Conservation* **6**, 69–74, 1981.
10. E. H. Pryde. Vegetable oils as fuel alternatives, *American Society of Agricultural Engineers* **4**, 101, 1984.
11. J. J. Bruwer, B. D. Boshoff, F. J. C Hugo, L. M. DuPlessis, J. Fulsand, and C. Hawkins. Using unmodified vegetable oils as a diesel fuel extender, *American Society of Agricultural Engineers* **32**(5), 1503, 2002.
12. R. A. Baranescu and J. J. Lusco. Performance, durability and low temperature evaluation of sunflower oil as a diesel fuel extender, *American Society of Agricultural Engineers* **4**, 312, 1982.
13. L. E. Wagner and C. L. Peterson. Performance of winter rape based fuel mixtures in diesel engines, *American Society of Agricultural Engineers* **4**, 329–336, 1982.
14. S. C. Borgelt and F. D. Harris. Endurance tests using soybean oil–diesel fuel mixture to fuel small pre-combustion chamber engines, *American Society of Agricultural Engineers* **82**(4), 364, 1982.
15. N. J. Barsic and A. L. Humke. Vegetable oils: Diesel fuel supplements? *Automotive Engineering* **89**(4), 37–41, 1981.
16. M. Ziejewski and K. R. Kaufman. Endurance tests of a sunflower oil/diesel fuel blends, *Journal of American Oil Chemists' Society* **61**(10), 1620, 1984.
17. T. K. Bhattacharya, T. N. Mishra, B. Singh, and D. P Darmora. Potential of some alternative fuels for I.C. engines, In: *20th Annual Convention of Indian Society of Agricultural Engineers,* 1983, Pantnagar, India, Paper No. 83–4723.
18. W. D. Samson, C. G. Vidrine, and J. W. D. Robbins. Chinese tallow seed oil as a diesel fuel extender, *Transactions of the American Society of Agricultural Engineers* **28**(5), 1406, 1983.
19. P. D. Dunn and E. D. I. H. Perera. Rubber seed oil for diesel oil in Sri Lanka, In: *Proceedings of the International Conference on Biomass,* March 25–29, 1985, Venice, Italy, p. 1172.
20. B. S. Samga. Use of Honni oil as an alternative fuel for C.I. engine, In: *Proceedings of the 9th National Conference on IC Engines and Combustion,* 1986, Dehradun, India, Paper No. A-20, 1.
21. D. L. Auld, B. L. Bettis, and C. L. Peterson. Production and fuel characteristics of vegetable oil from oilseed crops in the Pacific Northwest, *American Society of Agricultural Engineers* **4**(82), 92–100, 1982.

22. T. W. Ryan, L. G. Dodge, and T. J. Callahan. The effects of vegetable oil properties on injection and combustion in two different diesel engines, *Journal of American Oil Chemists' Society* **61**(10), 1610, 1984.
23. H. B. Mathur and L. M. Das. Utilization of nonedible wild oil as diesel engine fuel, In: *Proceedings of Bioenergy Society*, 1985, Hyderabad, India.
24. C. E. Goering, A. W. Schwab, R. M. Campion, and E. H. Pryde. Evaluation of soybean oil–aqueous ethanol microemulsions for diesel engines, *Journal of American Society of Agricultural Engineers* **4**(82), 279–286, 1982.
25. A. Nag, S. Haldar, and B. B. Ghosh. Studies on the Comparison of Performance and Emission Characteristics of a Diesel Engine Using Three Degummed Non-Edible Vegetable Oils (Unpublished Data).
26. F. Ma and M. A. Hanna. Biodiesel production: A review, *Bioresource Technology* **70**, 1, 1999.
27. D. Noureddini and D. Zhu. Kinetics of transesterification of soybean oil, *Journal of American Oil Chemists' Society* **74**, 1457–1463, 1997.
28. R. O. Feuge and T. Gros. Modification of vegetable oils. VII. Alkali-catalyzed inter-esterification of peanut oil with ethanol, *Journal of American Oil Chemists' Society* **26**, 97–102, 1949.
29. Y. Kawahara and T. Ono. US Patent No. 4,164,506, 1979.
30. K. J. Harrington and C. D'Arcy-Evans. Transesterification in situ of sunflower seed oil, *Industrial & Engineering Chemistry Product Research and Development* **24**, 314, 1985.
31. R. Stern, G. Hillion, P. Gateau, and J. C. Gulbet. Energy from biomass, In: *Proceedings of the International Conference on Biomass*, March 25–29, 1985, Venice, Italy.
32. A. W. Schwab, M. O. Bagby, and B. Freedman. Preparation and properties of diesel fuels from vegetable oils, *Fuel* **66**, 1372–1378, 1987.
33. K. Liu. Preparation of fatty acid methyl esters for gas-chromatographic analysis of lipids in biological materials, *Journal of American Oil Chemists' Society* **71**, 1179–1187, 1994.
34. J. M. Goff, S. N. Bauer, S. Lopes, W. R. Sutterlin, and J. G. Suppes. Acid-catalyzed alco-holysis of soybean oil, *Journal of American Oil Chemists' Society* **81**, 415–420, 2004.
35. L. C. Meher, S. N. Naik, S. N. and L. M. Das, L. M., Methonolysis of *Pongamia Pinnata* (Karanja) oil for production of biodiesel, *Journal of Scientific & Industrial Research* **63**, 913, 2004.
36. A. Isigigur, F. Karaosmanoglu, and H. A. Aksoy. Methyl ester from safflower seed oil of Turkish origin as a biofuel for diesel engines, *Applied Biochemistry and Biotechnology* **45**, 103, 1994.
37. D. Darnoko and M. Cheryan. Continuous production of palm methyl esters, *Journal of American Oil Chemists' Society* **77**(12), 1269–1272, 2000.
38. G. Knothe, R. O. Dunn, and M. O. Bagby. The use of vegetable oils and their deriva-tives as alternative diesel fuels, In: *Fuels and Chemicals from Biomass,* Saha, B. C. and Woodward, J. (Eds.), Washington, DC: American Chemical Society, pp.172–208, 1977.
39. B. Freedman, E. M. Pryde, and T. L. Mounts. Variables affecting the yields of fatty esters from transesterified vegetable oils, *Journal of American Oil Chemists' Society* **61**, 1638–1643, 1984.
40. H. P. Kruclen, H. C. A. VanBeek, E. Vander Drift, and G. Spruiji. *Proceedings of the International Conference on Biomass*, March 25–29, 1985, Venice, Italy, p. 1069.
41. G. Gelbard, O. Bres, R. M. Vargas, E. Vielfaure, and U. F. Schuchardt. Optimization of alkali-catalyzed transesterification of *Pongamia pinnata* oil for production of biodiesel, *Journal of American Oil Chemists' Society* **72**, 1239, 1995.
42. A. Chadha, K. S. Karmee, P. Mahesh, and R. Ravi. Study of the kinetics of base cat-alyzed transesterification of monoglycerides from *Pongamia oil*, *Journal of American Oil Chemists' Society* **81**, 425–430, 2004.
43. A. Nag. Alcoholysis of vegetable oil catalyzed by an isozyme of *Candida rugosa* lipase for production of fatty acid esters, *Indian Journal of Biotechnology* **5**, 175–178, 2006.
44. M. Kaieda, T. Samukawa, T. Matsumoto, K. Ban, and A. Kondo. Biodiesel fuel pro-duction from plant oil catalyzed by *Rhizopus oryzae* lipase in a water-containing system without an organic solvent, *Journal of Bioscience and Bioengineering* **88**(6), 627, 1999.

45. M. Oda, M. Kaieda, S. Hama, H. Yamaji, A. Kondo, E. Izumoto, and H. Fukuda. Facilitatory effect of immobilized lipase-producing *Rhizopus oryzae* cells on acyl migration in biodiesel-fuel production, *Journal of Biochemical Engineering* **23**, 45, 2005.
46. O. Köse, M. Tüter, and A. H. Aksoy. Immobilized *Candida antarctica* lipase-catalyzed alcoholysis of cotton seed oil in a solvent-free medium, *Bioresource Technology* **83**(2), 125, 2002.
47. H. Noureddini, X. Gao, and R. S. Philkana. Immobilized *Pseudomonas cepacia* lipase for biodiesel fuel production from soybean oil, *Bioresource Technology* **96**, 769, 2005.
48. Y. Shimada, Y. Watanabe, A. Sugihara, and Y. Tominaga. Enzymatic alcoholysis for biodiesel fuel production and application of the reaction to oil processing, *Journal of Molecular Catalysis B: Enzymatic* **17**, 133, 2002.
49. V. Dossat, D. Combes, and A. Marty. Continuous enzymatic transesterification of high oleic sunflower oil in a parked bed reactor: Influence of the glycerol production, *Enzyme and Microbial Technology* **25**, 194, 1999.
50. D. Royon, M. Daza. G. Ellenriedera, and S. Locatellia. Enzymatic production of biodiesel from cotton seed oil using *t*-butanol as a solve, *Bioresource Technology* **98**, 648, 2007.
51. Y. Watanabe, Y. Shimada, A. Sugihara, and Y. Tominaga. Conversion of degummed soybean oil to biodiesel fuel with immobilized *Candida antarctica* lipase, *Journal of Molecular Catalysis B: Enzymatic* **17**, 151, 2002.
52. W. Du, Y. Xu, D. Liu, and J. Zeng. *Journal of Molecular Catalysis B: Enzymatic* **30**, 125, 2004.
53. C. S. Hawkins, J. Fuls, and F. J. C. Hugo. Society of Automotive Engineers Trans (Index Abstracts), Paper No. 831356, 92, 191.
54. E. F. Fort, and P. N. Blumberg. Society of Automotive Engineers Trans (Index Abstracts), Paper No. 820317, 91, 63, 1982.
55. M. Ziejewski and K. R. Kaufman. Laboratory endurance test of a sunflower oil blend in a diesel engine, *Journal of American Oil Chemists' Society* **60**, 1567, 1983.
56. A. R. Tahir. Sunflower oil: An anticipated diesel fuel alternative, *Agricultural Mechanization in Asia, Africa, and Latin America* **16**(3), 59, 1985.
57. R. W. Pryor, M. A. Hanna, J. L. Schinstock, and L. L. Bashford. Soybean oil fuel in a small diesel engine, *Transactions of the American Society of Agricultural Engineers* **26**(2), 333, 1983.
58. S. J. Clark, L. Wagner, M. D. Schrock, and P. G. Piennaer. Methyl and ethyl soybean esters as renewable fuels for diesel engines, *Journal of American Oil Chemists' Society* **61**(10), 1632, 1984.
59. F. Laforgia, and V. Ardito. Biodiesel fueled IDI engines: Performances, emissions and heat release investigation, *Bioresource Technology* **51**, 53, 1995.
60. G. H. Pischinger, R. W. Siekmann, A. M. Falcon, and F. R. Fernandes. Methyl Esters of Plant Oils as Diesels Fuels Either Straight or in Blends. Vegetable Oil Fuels, 1982, American Society of Agricultural Engineers, St. Joseph, MI, Publication No. 4–82, p. 101.
61. Y. Ali, M. A. Hanna, and L. L. Leviticus. Emissions and power characteristics of diesel engines on methyl soyate and diesel fuel blends, *Bioresource Technology* **52**, 185, 1995.
62. R. E. H. Sims. Tallow esters as an alternative diesel fuel, *Transactions of the American Society of Agricultural Engineers* **28**(3), 716, 1985.
63. D. W. Richardson, R. J. Joyee, T. A. Lister, and D. F. S. Natusch. In: *Proceedings of the International Conference on Biomass*, March 25–29, 1985, Venice, Italy, p. 735.
64. M. Mittelbach and P. Tritthard. Diesel fuel derived from vegetable oils: Emission tests using methyl esters of used frying oil, *Journal of American Oil Chemists' Society* **65**, 707, 1988.

Ethanol and Methanol as Fuels in Internal Combustion Engines

B. B. Ghosh and Ahindra Nag

7.1 Introduction

The increasing industrialization and motorization of the world has led to a steep rise in the demand of petroleum products. Petroleum-based fuels are stored fuels in the earth. There are limited reserves of these stored fuels, and they are irreplaceable. Figure 7.1 shows the difference in demand and supply of petroleum products, and how this depletion will create a problem before the world within a decade or two.

Geologists throughout the world have been searching for further deposits. Although the present reserves seem vast, the accelerating consumption is challenging the world to create new types of fuels to replace the conventional ones. New oil reserves appear to grow arithmetically while consumption is growing geometrically. Under this situation, when consumption overtakes discovery, the world will be heading toward an industrial disaster.

Apart from the problems of fast-vanishing reserves and the irreplaceable nature of petroleum fuels, another important aspect of their use is the extent and nature of environmental pollution caused by combustion in vehicular engines. Petroleum-fueled vehicles discharge significant amounts of pollutants like CO, HC, NO_x, soot, lead compounds, and aldehydes.

A light-vehicular engine (car engine) discharges 1–2 kg of pollutants a day, and a heavy automobile discharges 660 kg of CO a year. CO is highly toxic, and exposure for a couple of hours to concentrations of 30 ppm can cause measurable impairments to physiological functions.

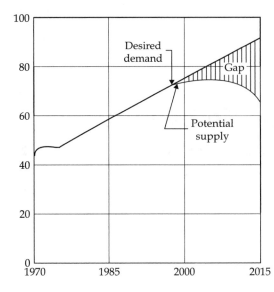

Figure 7.1 Difference in demand and supply of petroleum products.

Oxides of nitrogen and unburned hydrocarbons from exhausts cause environmental fouling by forming photochemical smog. Their interaction involves the formation of certain formaldehydes, peroxides, and peroxyacylnitrate, which cause eye and skin irritation, plant damage, and reduced visibility. Present day leaded gasoline contains lead compounds. Lead coming out with the exhaust finds its way into the human body, and causes brain damage in infants and children.

Vehicular exhaust fouling of the environment has already become a serious problem in Western countries and is a growing menace in developing countries like India [1]. They exhaust huge quantities of harmful pollutants in urban areas. Everyday, vehicles running in Delhi discharge about 240 tons of CO, 30 tons of HC, 20 tons of NO_x, and 2 tons of SO_2. The disastrous effect of these pollutants on human health, animal and plant life, and property are well known.

In view of these problems, attempts must be made to develop technology to produce alternative, clean-burning synthetic fuels. These fuels should be renewable, should perform well in the engine, and their potential for environmental pollution should be quite low.

Various fuels have been considered as substitutes for petroleum fuels used in automobiles. The most prominent of these include ethanol, methanol, NH_3, H_2, and natural gases [2]. The suitability of each of these fuels for internal combustion (IC) engines used in automobiles has been under investigation throughout the world. A few of them are already in use in different countries. This chapter introduces different types of unconventional fuel such as ethanol and methanol, their burning properties when used in IC engines, their performance characteristics compared with conventional engines, the modifications required in the engine if used in practice, and their environmental pollution characteristics.

7.2 Alcohols as Substitute Fuels for IC Engines

Due to the global energy crisis and continuous increase in petroleum prices, scientists have been in search of new fuels to replace conventional fuels that are used in IC engines. Among all the fuels, alcohols, which can be produced from sugarcane waste and many other agricultural products, are considered the most promising fuels for the future. There are two types of alcohols: ethanol (C_2H_5OH) and methanol (CH_3OH). Many other agricultural products (renewable sources) also have a vast potential for alcohol production, and it is necessary to tap this source to the maximum level in national interest. The use of alcohol as a motor fuel is itself not a new idea. Nicolas Otto, the pioneering German engine designer, suggested it as early as 1895. But, as long as crude oil was plentiful and inexpensive, petroleum gasoline was the most economical fuel for the IC engine.

Due to the global energy crisis, many countries that used to export molasses to be used as cattle feed are now setting up distilleries to manufacture ethanol.

7.2.1 Ethanol as an alternative fuel

Ethanol (ethyl alcohol) as a transport fuel has attracted a lot of attention because it is seen as a relatively cheap nonpetroleum-based fuel. It is produced to a large extent from biomass, which aids agricultural economies by creating a stable market. Ethanol, being a pure compound, has a fixed set of physical as well as chemical properties. This is in contrast to petrol and diesel, which are mixtures of hydrocarbons [3].

The use of alcohol in spark ignition (SI) engines began in 1954 in countries like the United States, Germany, and France. During World Wars I and II, gasoline shortages occurred in France and Germany, and alcohol was used in all types of vehicles, including military planes. Nowadays, it is used with gasoline (a mixture) in the United States and has become a major fuel in Brazil.

Ethyl alcohol can be produced by fermentation of vegetables and plant materials. But in countries like India, ethanol is a strong candidate since they possess the agricultural resources for the production of ethyl alcohol. It is a more attractive fuel for India because the productive capacity from sugarcane crops is high, of the order 1345 L/ha. Earlier, this fuel was not used in automobiles due to low energy density, high production cost, and corrosion. The current shortage of gasoline has made it necessary to substitute ethanol as fuel in SI engines.

TABLE 7.1   Comparative Properties of Ethanol with Petrol and Diesel

Sr no.	Property	Petrol	Diesel	Ethanol
1.	Specific gravity (at 15°C)	0.73	0.82	0.79
2.	Boiling point (°C)	30–225	190–280	78.3
3.	Specific heat (MJ/kg)	43.5	43.0	27.0
4.	Heat of vaporization (kJ/kg)	400	600	900
5.	Octane number (Research)	91–100	NA	NA
6.	Cetane number	Below 15	40–60	Below 15

Any new fuel that is going to be introduced should be evaluated from the aspect of availability, renewability, safety, and cost adaptability to the existing engines' performance, economy, and finally emission. A massive research effort has been put into the study and analysis of all these aspects for ethanol, which is now an established, viable alternative fuel for IC engines. The comparative properties of ethanol with petrol and diesel are shown in Table 7.1.

7.2.2 Production of ethanol

Ethanol is the most appropriate fuel for India to replace petrol, and the utmost of efforts have been made to increase alcohol production in the country. India is in an extremely happy position in this regard as it is the world's largest producer of sugarcane, a major source of alcohol. India topped the world in sugar production with 181 Mton (in 1978), followed by Brazil (130 Mton) and Cuba (67 Mton).

Alcohol is derived not directly from sugarcane but molasses–sugarcane by-products. All starch-rich plants like maize, tapioca, and potato can be used to produce alcohol; cellulosic waste materials can also be used. Production of ethanol from biomass involves fermentation and distillation of crops. India has a vast potential to produce ethanol, and only 2.5% of the country's irrigated land is used to produce sugarcane. This can be raised to a much higher level without adversely affecting the production of food-bearing crops.

At present, Brazil is the only country that produces fuel alcohol on a large scale from agricultural products (mainly sugarcane). Other countries, especially those with an substantial agricultural surpluses, such as the United States and Canada, are also bound to enter into this field of so-called energy forming. The area of land required is substantial. A medium-sized car with an annual run of 15,000 km needs 2000 L of ethanol. To produce this amount, the crop areas required are given in Table 7.2. To provide enough sugar beet alcohol to fuel 20 million cars in Germany requires half the area of the entire country.

TABLE 7.2 Crops Area Required for Growth

Crop	Sugarcane	Sweet sorghum	Sugar beet	Cassava	Potatoes	Wheat
Area (ha)	0.49	0.38	0.5	1.43	1.2	2.52

Sugarcane. The present method adopted to obtain alcohol for energy purposes requires three stages: (1) extracting the juice from sugarcane, (2) fermentation of the juice, and (3) distillation into 90–95% alcohol.

Molasses. The black residue remaining after the sugar is extracted from sugarcane is called molasses. It contains mostly invert sugars and some sucrose. This sucrose also undergoes hydrolysis to produce invert sugar by a catalytic action of acids in molasses.

$$C_{12}H_{22}O_{11} + H_2O \rightarrow C_6H_{12}O_{11} \text{ (D-Glucose)} + C_6H_{12}O_6 \text{ (D-Fructose)}$$

This mixture product is not crystallizable. Yeast organisms in the presence of oxygen oxidize sugars into CO_2 and H_2O and convert sucrose mostly into ethyl alcohol.

$$C_6H_{12}O_6 \rightarrow 2C_2H_5OH + 2CO_2$$

Process adopted. Molasses is mixed with water so that the concentration of sugar in it is 10–18% (optimum is 12%). If the concentration is high, more alcohol may be produced and may kill the yeast. Then, a selected strain of yeast is added (it should not contain any wild yeast). For some nutrient substances like ammonium and phosphates, the pH value is kept between 4 and 5, which favors the growth of yeast organisms. H_2SO_4 is used for lowering pH. The temperature of the mixture is kept at 15–25°C. The fermentation takes place as follows:

1. First, the yeast cells multiply at an optimum temperature (30°C).

2. Rapid fermentation takes place at the boiling temperature, and oxygen is given off. The optimum temperature (50°C) is maintained, and the process is continued for 20–30 h.

3. The fermentation rate is reduced, and alcohol is produced slowly. Total time for fermentation is 36–48 h, depending upon the temperature and sugar content. Last, the formed ethanol is distilled.

Starch. In this process, starchy materials are first converted into fermentable sugars. This is done by enzymatic conversion (by means of malt process) or by acid hydrolysis.

$$\text{Starch} \rightarrow C_{12}H_{22}O_{11} \text{ (Maltose)} + C_6H_{12}O_6 \text{ (Dextrose)}$$

Malt process.   Malt is prepared by germination of barley grains to produce required enzymes. The grain is ground and steam cooked at 100–150°C to break the cell wall of starch. For every 25 kg of grain, 100 L of water is added. Then the formed mass is cooled to 60–70°C and taken to large vessels where malt is added within 2 h and 60–70% of the stock is converted into maltose. Converted mash is cooled to a fermenting temperature of 20–25°C. pH is adjusted and fermentation is affected, producing ethanol.

Acid hydrolysis.   This process involves treatment with concentrated sulfuric or hydrochloric acid at pH 2–3 and 10–20 kg pressure in an autoclave to make sugar and then conversion of sugar to alcohol by yeast.

Cellulose material.

Wood.   Cellulose from wood is hydrolyzed into simple sugars by using diluted acid at a high temperature or concentrated acid at a low temperature. Similarly, cellulosic agricultural waste and straws can be used in place of wood.

Sulfite waste liquor from paper manufacture.   Waste liquor contains 2–3.5% of sugar, out of which 65% is fermentable into alcohol. Before fermentation, all acids in the liquor are removed by adding calcium. Then fermentation is carried out by special yeasts. Generally, 1% of liquor is converted into alcohol.

Hydrocarbon gases.

Hydration of ethylene.   Conversion of ethylene to ethyl alcohol can be carried out with high yield by first treating ethylene with H_2SO_4, forming ethyl hydrogen sulphate and diethyl sulfate, as given by the following reactions:

$$C_2H_5HSO_4 \rightarrow (C_2H_5)_2SO_4$$

$$2C_2H_4 + H_2SO_4 \rightarrow (C_2H_5)_2SO_4$$

These products, ethyl sulfuric acid and diethyl sulfate, when treated with water give ethanol as per the following reactions:

$$C_2H_5HSO_4 + H_2O \rightarrow C_2H_5OH + H_2SO_4$$

$$(C_2H_5)_2SO_4 + 2H_2O \rightarrow 2C_2H_5OH + H_2SO_4$$

Direct hydration.   Ethanol is also formed as per the following chemical reaction:

$$C_2H_4 + H_2O \rightarrow C_2H_5OH$$

This type of conversion is very small as the reaction is exothermic; it is not a suitable method for mass production. The corn is first ground, then mixed with water and enzymes, and cooked at 150°C to convert starch to sugar. The mixture is then cooled and sent to fermentation tanks, where yeast is added and the sugar is allowed to ferment into ethanol. After 60 h in the tanks, the mixture is sent to distillation columns, where ethanol is evaporated out, condensed, and mixed with unleaded gasoline to form gasohol, which contains 90% gasoline and 10% ethanol.

Tapioca materials.   Tapioca is available in plenty in Asia, the United States, central Europe, and Africa. Its production can be increased through modern cultivation techniques. The process consists of converting the tapioca flour into fermentation sugars with enzymes prior to fermentation with yeast. Modern technology uses α-amyl glycosidase, one of two enzymes required in the process, and then saccharification of the material into alcohol by using yeast.

Anhydrous alcohol from vegetable wastes.   The Philippines has embarked on an "alcogas program" to produce its own anhydrous alcohol from local vegetable wastes for blending with petrol. The program is currently based on sugarcane juice and molasses, but it plans to diversify by using other raw materials. In the basic process, cellulose conversion begins with the pretreatment of the raw materials, which may include coffee hulls, rice straw, grass—even sawmill wastes. Enzymes then take over by converting the feedstock into a sugary liquid that is fermented and finally distilled into anhydrous alcohol. After distillation, waste residues can be evaporated into syrup to feed animals, while unconverted cellulose is used as the primary fuel for the plant. If the Philippines could engineer a breakthrough in this area, its agricultural and forestry wastes could supply energy equivalent to 9720 mL of oil annually. In the years to come, this new energy source could make a significant economic impact on a country that depends on imports of crude oil for 95% of its energy.

Manioc.   As oil prices continue to rise, more and more work is being done on alternatives. Manioc is one such staple crop in many tropical lands. Brazil has planned to use manioc in its ethanol production plants, aiming to make 35,000 bbl a day from 400×10^3 ha of manioc plantation. Conversion of manioc to ethanol is somewhat more complex than is the case with sugarcane. The raw material has to be turned into sugar by fermentation. This first step requires the use of enzymes. Danish Co. has developed the necessary heat-resistant enzymes in a pilot plant in Brazil.

Manioc does not grow in higher temperature zones; so scientists have turned to other plants, and there is work being done in Sweden that is in an advanced stage. They have developed fast-growing poplars and willows. Their yield is 30 ton/ha, which is equal to 12 tons of fuel oil.

It is estimated that 1000×10^3 ha planted with such trees can provide 10% of Sweden's electricity. Also in Sweden, work has been carried out on the common reed, and the estimated yield is 10 ton/(yr · ha), which is equal to 4.5 tons of oil. Sweden has plans to have 100×10^3 ha of reeds. Brazil's program of ethanol from sugarcane and manioc may employ 200×10^3 people and save $1600 million each year in foreign exchange.

7.3   Distillation of Alcohol

If a mixture of water and alcohol is boiled, the percentage of alcohol to water is greater in vapor than in liquid. Therefore, by repeated distillation and condensation, the alcoholic strength of the distillate can be increased until it contains 97.6% alcohol. There are different methods of distillation, but they are not discussed here, as ethanol production is our prime concern.

7.4   Properties of Ethanol and Methanol

Both ethanol and methanol, as listed in Table 7.3, have high knock resistance (as the octane numbers are 89 and 92, against 85 for gasoline), wide ignition limit, high latent heat of vaporization, and nearly

TABLE 7.3   Important Alcohol Properties

Sr no.	Property	Gasoline C_8H_{18} isooctane	Ethyl alcohol	Methyl alcohol
1.	Molecular weight (g)	114.2	46	32
2.	Boiling point at 1 bar (°C)	43–170	78	66
3.	Freezing point (°C)	−107.4	117.2	−161.8
4.	Specific gravity (150°C)	0.72–0.75	0.79	0.79
5.	Latent heat (kJ/kg)	400	900	1110
6.	Viscosity (centipoise)	0.503	0.60	0.596
7.	Stoichiometric A:F (ratio)	14.6	9	6.45
8.	Mixture heating value (kJ/kg) (for stoicmixture)	2930	2970	3070
9.	Ignition limits (A/F)	8–19	3.5–17	2.15–2.8
10.	Self-ignition temperature	335	557	574
11.	Octane number			
	a. Research	80–90	111	112
	b. Motor	85	92	91
12.	Cetane number	15	8	3
13.	Lower CV (kJ/kg)	44,100	26,880	19,740
14.	Vapor pressure at 38°G (bar)	0.48–1	0.17	0.313
15.	Flame speed (m/sec)	0.43	—	0.76
16.	Autoignition temperature (°C)	222	—	467

the same specific gravity. All those properties are of great advantage if used in SI engines. Some important advantages of alcohol-fueled engines compared with gasoline engines are listed below:

1. The alcohols (both) have higher heat of vaporization. As the liquid fuel evaporates into the air stream being charged to the engine, a higher heat of vaporization cools the air, allowing more mass to be drawn into the cylinder. This increases the power produced from the given engine size. High latent heat of vaporization leads to higher volumetric efficiency and provides good internal cooling.

2. The high octane number of alcohols compared to petrol means higher compression ratios can be used, which results in higher engine efficiency and higher power from the engine.

3. Ethanol burns faster than petrol, allowing more uniform and efficient torque development. Both alcohols have wider flammability limits, which results into a rich air–fuel (A:F) ratio being used when needed to maximize power by injecting more fuel per cycle.

4. Alcohols also have lower exhaust emissions than gasoline engines except for aldehydes. Both alcohols have lower carbon–hydrogen ratio than petrol and diesel, and produce less CO_2. For the same power output, CO_2 produced by an ethanol-fired engine is about 80% of the petrol engine. Because of high heat of vaporization, the fuels burn at lower flame temperatures than petrol, forming less NO_x. The CO percentage in both cases (alcohol and petrol) remains more or less the same.

5. Contamination of water in alcohols is less dangerous than petrol or diesel because alcohols are less toxic to humans and have a recognizable taste.

6. The alcohols can also be blended with gasoline to form the so-called gasohol (80% petrol and 20% alcohol), which is widely used in the United States.

7. Ethyl alcohol as a fuel offers great safety due to its low degree of volatility and higher flash point (17°C).

8. The heating value of alcohol is 60% of that of petrol (60% only), and it shows equally good thermal efficiency and lower fuel consumption, because the air required for petrol and alcohol is in the ratio of 15:9 by weight, which is the same as their calorific value, i.e., the same heat is developed per cylinder charge in petrol and alcohol engines. The power per unit volume of cylinder for petrol, ethanol, and methanol are closely similar.

9. In many hot-climate countries, more precautions are often taken for the use of more volatile spirit-based fuels, while alcohol is perfectly safe in the hottest climate.

10. The major problem faced with ethanol is corrosion; special metals should be used for the engine parts to avoid corrosion.

Alcohols are clean-burning, renewable alternative fuels that can come to our rescue to meet the duel challenge of vehicular fuel oil scarcity and fouling of the environment by exhaust emissions.

Alcohols inherently make very poor diesel engine fuels as their cetane number is considerably lower. They can be used in dual-fuel engines or with assisted ignition in diesel engine. In the dual-fuel mode, alcohol is inducted along with air, compressed, and then ignited by a pilot spray of diesel oil.

7.5 Use of Blends

Alcohol can be used as a blend with gasoline as this has the advantage that the existing engines need not be modified and tetra-ethyl lead (TEL) can be eliminated from gasoline, due to the octane-enhancing quality of alcohol. If the engine is to be operated using only pure alcohol, then some major modifications are required in the engine and fuel system, as listed below:

1. Both alcohols and blends with gasoline are corrosive to many of the engine materials. These materials have to be changed.

2. Adjustment of the carburetor and fuel injection need to be made to compensate for the leaning effect.

3. Change in the fuel pump and circulation system need to be made to avoid vapor lock, as the methanol vaporization rate is very high.

4. Introduction of high energy ignition system with lean mixture.

5. Increase in compression ratio to make better antiknock properties of the fuel.

6. Addition of detergent and volatile primers to reduce engine deposits and assist in cold starting.

7. Use of cooler-running spark plugs to avoid preignition.

General properties of the blends are listed in Table 7.4. The volatility shown by the American Standard Testing Method (ASTM) distillation characteristics of petrol is a compromise between opposing factors to ensure good performance in petrol engines. This requires petrol to have a sufficiently lighter reaction and a 10% distillation temperature in order to start the engine as well as warm up, but the temperature should

TABLE 7.4   Evaluation of Ethanol and Gasohol against Petrol

Characteristics	BIS specification for petrol	Petrol	Ethyl alcohol	Petrol and Ethyl Alcohol (Gasohols)			
				95%+ 5%	90%+ 10%	85%+ 15%	80%+ 20%
ASTM distillation							
Initial boiling point, °C	....	55	78	55	50	48	46
10% volume	70 (max)	64	....	59	56	57	57
50% volume	125 (max)	92	....	95	73	70	70
90% volume	180 (max)	128	....	145	127	130	125
Final boiling, °C	215 (max)	143	....	....	147	156	156
Gum residue, mg/100	4	22	55	51	91	131	180
Aniline point, °C	....	44	30	40	35	32	30
Specific gravity	....	0.720	0.7966	0.7230	0.722	0.7289	0.733

not be so low that vapor-locking takes place and stops the engine due to the nonsupply of fuel. As far as volatility is concerned, ethanol–petrol blends are as good as petrol, if not better. Also gum resistance is greater than that of petrol. Aniline points for blends are lower, which indicates more aromatic content than petrol, due to the adding of ethanol to petrol, which helps to improve the octane number marginally. If a small quantity of water is introduced into a gasoline–alcohol blend, phase separation takes place, with gasoline–content in the upper phase and alcohol in the lower. This separation produces some undesirable effects. The alcohol–water mixture tends to pick up sediment and stall the engine on reaching the carburetor [4]. To improve the water tolerance of the blend, benzene and heptanes are added.

Since 1979, gasohol has been sold at 500 filling stations in the midwestern United States, where the corn from which alcohol is commonly made is abundant. This blend yields about the same mileage as unleaded gasoline and even offers an ever renewable source of energy. Moreover, if this blend were to replace gasoline, it could cut as much as 10% of the nation's oil imports, which totalled $40 billion in 1979. This fuel has a good future in wealthy countries. The blends have some important advantages over pure ethanol, as listed below:

1. The starting difficulty can be removed.

2. There is no abnormal corrosion compared with pure ethanol.

3. Lubrication in a petrol–alcohol blend is more or less the same.

4. Some benzene is added to prevent separation of the layers of petrol and alcohol.

If blends are used, some minor modifications in the engine are required, as listed below:

1. The carburetor jet should be increased to increase the flow 1.56 times that of petrol.

2. The float has to be weighted down to correct levels due to higher specific gravity.

3. The air inlet should be modified to get less air as blends require less air for complete combustion than petrol.

4. Specific arrangement of heating the carburetor and intake manifold should be provided as lower vapor pressure of alcohol makes the starting difficult below 70°C.

7.6   Performance of Engine Using Ethanol

The effect of speed on power output, brake specific fuel consumption (BSFC), and thermal efficiency of an engine using ethanol is compared with gasoline engine, is shown in Figs. 7.2 through 7.5.

The observations are listed below:

1. The power output of the ethanol engine is higher, compared to a gasoline engine at all speeds.

2. The BSFC is improved with an ethanol engine, compared to a petrol engine.

3. The maximum thermal efficiency of an ethanol engine is higher than that of a petrol engine. The efficiency curve of an ethanol engine is flat for a wide range of speeds, which indicates that the partial-load efficiency is much better, compared with a petrol engine.

4. The engine torque is considerably higher for ethanol as compared to a petrol engine.

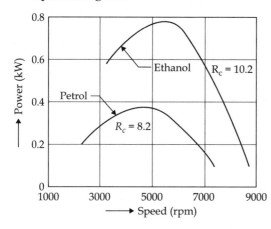

Figure 7.2 Effect of speed on power at different compression ratios.

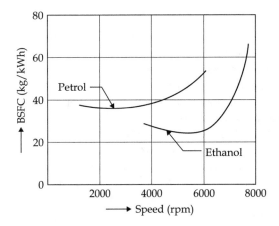

Figure 7.3 Effect of speed on BSFC (brake specific fuel consumption).

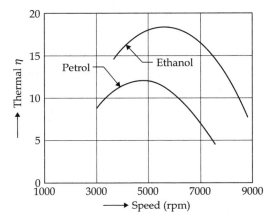

Figure 7.4 Effect of speed on thermal efficiencies.

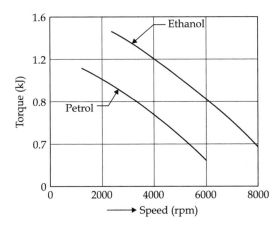

Figure 7.5 Effect of speed on the torque.

7.7 Alcohols in CI Engine

Although the physical and thermodynamic characteristics of alcohols do not make them particularly suitable for compression ignition (CI) engines, with certain modifications, however, they can also be used in CI engines. In heavy vehicles powered by CI engines, ethanol carburetion can be employed for bi-fuel operation of the engine with proportional savings in diesel oil. The various methods for using alcohols with diesel are fumigation, dual injection, and alcohol–diesel emulsions.

In a fumigation system the engine is fitted with a suitable carburetor and auxiliary ethanol tank. An ethanol-air mixture is carbureted during the induction stroke to provide 50% of the total energy of the cycle and the remaining energy is provided by diesel oil being injected in the conventional manner near the end of the compression stroke. The materials of a fuel tank and fuel system must be compatible with alcohol. The entire system can be used as a retrofit kit, as shown in Fig. 7.6.

Ghosh et al. [4] carried out an investigation on the performance of a tractor diesel engine with ethanol fumigation (see Figs. 7.7 and 7.8). The following observations were recorded:

1. The brake thermal efficiency decreases with an increase in ethanol fumigation rate at a constant engine speed.

2. The BSFC decreases with an increase in ethanol fumigation rate at a constant engine speed.

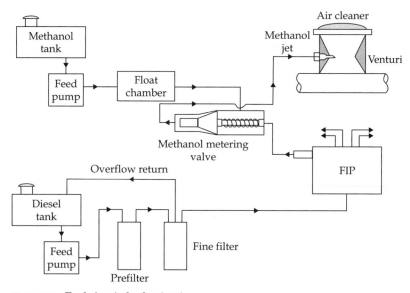

Figure 7.6 Fuel circuit for fumigation.

Figure 7.7 Experimental setup of ethanol fumigation.

3. The diesel substitution and the energy replacement increases with an increase in an ethanol fumigation rate at a constant engine speed.

4. The NO_x emission level and the exhaust gas temperature decreases with an increase in a ethanol fumigation rate at a constant engine speed.

5. The CO emission level increases with an increase in an ethanol fumigation rate at a constant engine speed.

6. The smoke level decreases with an increase in an ethanol fumigation rate at a constant engine speed.

7. The fumigation rate of 1.06 kg/h (40% diesel substitution) is optimal for good engine performance.

Figure 7.8 Ethanol fumigation nozzle.

Ethanol fumigation in diesel engines can play a major role in environmental air pollution control, and ethanol is a viable alternative fuel for diesel engines.

Ethanol is a very good SI engine fuel and a rather poor CI engine fuel. Ethanol has a high octane rating of 90 and a low cetane rating of 8, and will not self-ignite reasonably in most CI engines. Dehydrated ethanol is fumigated into the air stream in the intake manifold of a 42-hp tractor diesel engine to improve its self-ignition quality. The performance of the engine under dual-fuel (diesel and fumigated ethanol) operation is compared with diesel fuel operation at various speeds (800, 900, 1000, 1100 rpm), loads (0, 4, 8, 12, 16 kgf), and fumigation rates (0.00, 1.06, 1.45, 2.06 kg/h). Analysis of the results shows that ethanol fumigation has the advantages of reduction in BSFC, NO_x emission, and smoke level and the disadvantage of slight reduction in brake thermal efficiency. The fumigation rate of 1.06 kg/h (40% diesel substitution) is optimal for good engine performance.

It has been concluded that ethanol is a viable alternative fuel for diesel engines. A dramatic reduction in the NO_x and the smoke level suggests that fumigation, as an emission control technique in diesel engines, can play a vital role in environmental air pollution control on a farm.

In the dual-injection method, two injection systems are used, one for diesel and the other for alcohol. This method can replace a large percentage of diesel fuel. In this method, air is sucked and compressed, and then methanol is injected through a primary injector. To ignite this, a small amount of diesel is injected through a pilot injector. The relative injection timing of alcohol and diesel is an important aspect of the system.

As two injection systems are required, two injectors are required on the cylinder head, which limits the application of this method to large-bore engines. An additional pump, fuel tank, and fuel line are also required, making the system more complicated. But this method replaces 60% of diesel at a partial load and 90% at a full load, and provides higher thermal efficiency.

7.7.1   Alcohol–diesel fuel solution

This method is the easiest but requires anhydrous ethanol, because methanol has limited solubility. A maximum of 10% diesel can be substituted due to the lower solubility of methanol in diesel. No component changes; only adjustments of injection timing and fuel volume delivery are required to restore full power. Dodecanol is an effective surfactant for methanol–diesel fuel blends. Straight-run gasoline is an economical additive for ethanol–diesel blends.

Solubility of alcohols in diesel fuels is a function of (a) fuel temperature, (b) alcohol content, (c) water content, (d) specific gravity of diesel,

(e) wax content, and (f) hydrocarbon composition. Methanol solubility in diesel increases as the aromatic content goes up.

7.7.2 Alcohol–diesel fuel emulsions

Here, an emulsifier extends the water tolerance of alcohol–diesel blends. In general, equal volumes of alcohols and emulsifiers are required for suitable emulsions. No component changes, but injection volume and timing are adjusted for diesel fuel with alcohol then solutions, i.e., up to 35% diesel substitution is possible. Addition of ignition improvers, e.g., cyclohexanol nitrate, up to 1% helps increase the alcohol percentage up to 35% while maintaining a cetane rating at permissible levels. Cost of emulsifiers and poor low-temperature physical properties of emulsions limit the use of this technique. Stable emulsion requires the use of costly surfactants. Using higher-order alcohols improves the stability of blends at temperatures as low as $-20°C$.

7.7.3 Spark ignition

This technique replaces 100% diesel. The injection system can be retained as is or replaced by carburetion or port-type fuel injection. A spark plug is introduced in the combustion chamber, and the associated ignition system is added. High compression ratio and positive ignition result in smooth combustion, thereby improving thermal efficiency.

This approach is quite attractive as it uses the high latent heat of the vaporization of alcohols and their octane rating to good advantage. Power output is reduced due to lower heat content of alcohols. Changes in engine operability are not noticeable with alcohol-fired SI engines, relative to the same engines using diesel fuel due to their similar torque. The engines are as efficient as their diesel-fueled counterparts. In fact, huge torque is available at engine speeds below 1400 rpm, which increases engine flexibility and response in use. Converting an existing diesel fleet to an SI technique involves engine modification. Space at the appropriate place must be available for spark plugs in the cylinder head. Lubricants need to be added to alcohols to increase lubricity and prevent wear. Small amounts of cetane improvers may be added, but they are not required. It is not easy to switch between fuels after conversion to the SI technique.

7.7.4 Ignition improvers

Neat alcohols are used in diesel engines by increasing the cetane number sufficiently using ignition improvers. This technique saves the expense and complexity of engine component changes but adds the cost of ignition improvers. The cost of 10–20% ignition improvers is quite prohibitive.

The most effective ignition improvers are nitrogen-based compounds which can aggravate exhaust emissions of NO_x. Ethylene glycol nitrates have shown promising trends at 5% concentration.

Engines operating on cetane-enhanced alcohol need a few changes, e.g.,

- Injection volume and timing must be adjusted to obtain optimum performance.

- A large pump, fuel lines, and injectors are required to satisfy total fuel requirements of the engine for the desired output.

- A lubricant (generally castor oil used so far) is required to be added to alcohols using improvers.

7.8   Methanol as an Alternate Fuel

Methanol behaves much like petroleum, so it can be stored and shifted in the same manner. It is a more flexible fuel than hydrocarbon fuels, permitting wider variation from the ideal A:F ratio. It has relatively good lean combustion characteristics compared to hydrocarbon fuels. Its wider inflammability limits and higher flame speeds have shown higher thermal efficiency and less exhaust emissions, compared with petrol engines.

Methanol can be used directly or mixed with gasoline. Tests have shown improvements in fuel economy by 5–13%, decreases in CO emission by 14–70%, and reductions in exhaust temperature by 1–9%, with varying methanol in petrol from 5 to 30%. Depending on the gasoline–methanol mixture, some changes in fuel supply are essential. Simple modifications to the carburetor or fuel injection can allow methanol to replace petrol easily. Some important features of methanol as fuel are listed below:

1. The specific fuel consumption with methanol as fuel is 50% less than a petrol engine.

2. Exhaust CO and HC are decreased continuously with blends containing higher percentage of methanol. But exhaust aldehyde concentration shows the opposite trend.

3. Like ethanol, methanol can also be used as a supplementary fuel in heavy vehicles powered by CI engines with consequent savings in diesel oil and reduced exhaust pollution. No undue wear of engine components are encountered with methanol as a fuel, while engine peak power improves and smoke density and NO_x concentration in exhaust is reduced.

Phase separation, vapor lock, and low-temperature starting difficulties are the problems associated with the use of methanol or its blends as

IC engine fuels. Availability from indigenous sources, ease of handling, low emission, and high thermal efficiency obtainable with its use make methanol a logical alternative fuel for vehicular engines.

7.8.1 Production of methanol

Methanol can be produced from resources such as coal, natural gas, oil shell, and farm waste, which are abundant worldwide. But methanol from natural gas is unlikely to provide a large greenhouse benefit, not more than a 10% reduction in emissions with quite optimistic assumptions. It is not considered as a main raw material to produce methanol. For countries having vast reserves of coal but small oil deposits, methanol from coal can provide an indigenous substitute to oil. But this method has an adverse effect on greenhouse gases and is very expensive, requiring capital investments that can increase the price by 50%.

In India, there is an abundant production of sugarcane. The government can divert this feedstock to produce methanol. The production of methanol by using water and methane is shown in Fig. 7.9, and by using methane and a catalyst in Fig. 7.10.

Producing methanol from methane with the technology available today generally involves a two-step process. Methane is fuel reacted with water and heat to form carbon monoxide and hydrogen—together called synthesis gas. Synthesis gas is then catalytically converted to methanol. The second reaction unleashes a lot of heat, which must be removed from the reactor to preserve the activity of the temperature-sensitive catalyst. Efforts to improve methanol synthesis technology

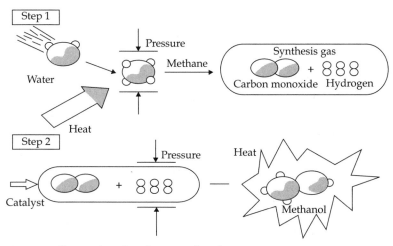

Figure 7.9 Conversion of methane to ethanol.

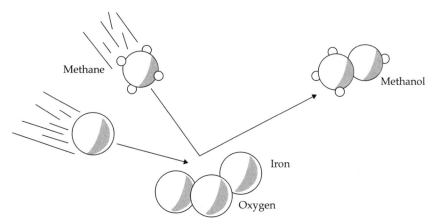

Figure 7.10 Production of methanol by using methane and a catalyst.

focus on sustaining the catalyst life and increasing reactor productivity. As a novel alternative to the two-step method, a chemical catalysis that mimics biological conversion of methane by enzymes is being developed. The iron-based catalyst captures a methane molecule, adds oxygen to it, and ejects it as a molecule of methanol. If this type of conversion could be performed on a commercial scale, it would eliminate the need to first reform methane into a synthesis gas, which is a costly, energy-intensive step. Conversion of coal to methanol is simpler and cheaper as compared to its liquefactions to gasoline.

Advantages of methanol.

1. 1% methanol in petrol is used to prevent freezing of fuel in winter.
2. Tertiary-butyl alcohol is used as an octane improving agent.
3. Because of the excellent antiknock characteristics of the fuel, it is very suitable for SI engines.
4. Isopropyl alcohol is used as an anti-icing agent in carburetos.
5. Addition of methanol causes a methanol–gasoline blend to evaporate at a much faster rate than pure gasoline below its boiling point (bp).
6. Due to an increase in emission levels of conventional fuels, the percentage of O_3 in the atmosphere is increasing. This increase in the O_3 in the atmosphere might cause biomedical and structural changes in the lungs which might cause chronic diseases. O_3 content of even between 0.14 and 0.16 ppm temporarily affects lung function if the person is exposed to it for 1–2 h. An annual crop yield is also reduced if exposed to O_3; some trees suffer injury to needles or leaves,

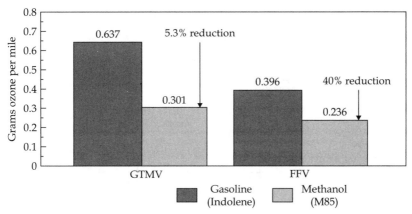

Figure 7.11 Effects of methanol on O_3 emission compared with petrol.

and lower productivity or even die. High content of O_3 has disturbed the natural ecological balance of species in national forests in California. The effects of methanol on O_3 emission as compared with petrol is shown in Fig. 7.11.

7.8.2 Emission

Methanol-fueled vehicles emit less CO_2 and other polluting gases compared to gasoline-fueled vehicles. Therefore, methanol use maintains good air quality. For a higher compression ratio compared to gasoline, a higher level of NO_x can be achieved. But low flame temperature and latent heat of vaporization tend to decrease NO_x emissions. The overall effect is a lower level of NO_x emissions.

7.8.3 Fuel system and cold starting

Methanol has high latent heat; therefore, some provision must be provided to heat the intake manifold, because cold starting problems are often caused by A:F vapor mixture being outside the flammability range. Specially, methanol in its pure form is much more inferior to petrol for cold starting. Cold starting more or less becomes impossible with methanol when the ambient temperature falls below system on chip (SOC). Figure 7.12 shows the modification that is provided to avoid the difficulty of cold starting. By preheating, methanol dissociates into CO + $2H_2$ to obtain gaseous H_2, which gives a broad flammability limit. While cranking the engine, a rich gaseous A:F mixture of methanol is collected near the spark plug, which enables good starting of the engine.

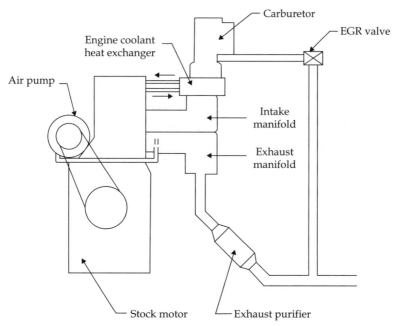

Figure 7.12   Modification provided to avoid the difficulty of cold starting.

7.8.4   Corrosion

Corrosion of the engine parts has been one of the main reasons for not using alcohols as fuels. The problem of corrosion is severe during starting and idling; but once the engine starts and gets heated, corrosion does not take place. Severe corrosion is noticed with Zn, Pb, Cu, Mg, and Al. This problem has been solved by using a methanol-resistant filter before the carburetor. Corrosion by methanol has been prevented by using the corrosion inhibitor LZ541 manufactured by M/S Lubrizol India. Being solvent, it swells or softens many parts of plastic or rubber commonly used for gaskets or floats in the carburetor. This is solved by using elastomers instead of rubber or plastic. American Motors' Gremlin model of 1970 has been used continuously for 9 years using pure methanol without facing any difficulty of corrosion. Two 1972 Plymouth Valiants have been used for 7 years: one using pure methanol and the other using a methanol blend without any difficulty. None of these vehicles has had a failure of engine components or fuel system components.

7.8.5   Toxicity of methanol

Methanol is more toxic as compared to petrol, which creates difficulty in its handling. The toxicity of methanol is reduced by adding chemical emetics.

7.8.6   Formaldehyde emission

The major problem with methanol is high levels of formaldehyde emission, which is negligible with conventional fuels. Formaldehyde emission levels with and without an electric heater are shown in Fig. 7.13. The level with an electric heater is considerably lower compared with its absence.

The performance characteristics compared with petrol engine are considered as brake thermal efficiency versus air fuel (A:F) ratio, the effect of speed power output and specific heat consumption. In addition, the performance characteristics also include the effect of A:F ratio on exhaust emission. The effects of A:F ratio and speed on brake power are shown in Figs. 7.14a and 14b. Another important characteristic is the effect of speed on volumetric efficiency, which is shown in Figs. 7.15 and 7.16.

Both alcohols, as well as their blends, are studied as alternative fuels for IC engines. The power can be increased from 6 to 10% with alcohols or their blends. The use of a leaner mixture provides more O_2, which reduces the emission. Because of the high heat of vaporization of these fuels compared to petrol, greater cooling of the inlet mixture occurs, which gives higher thermal efficiency, less specific heat consumption, and smooth operation. At higher speeds, the specific heat consumption is lower than that of petrol. Methanol dissociates in the engine cylinder forming H_2. This H_2 gas helps the mixture to burn quickly and increases the burning velocity, which brings about complete combustion and makes a leaner mixture more combustible. In a petrol engine, misfiring

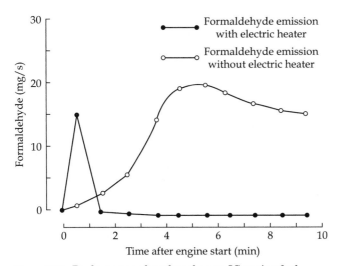

Figure 7.13   Performance of methanol as an IC engine fuel.

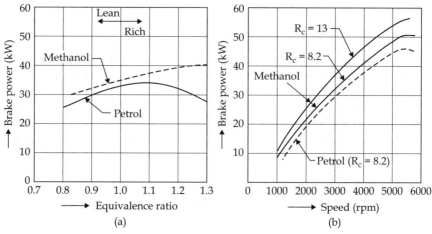

Figure 7.14   Effect of (a) equivalence ratio and (b) speed on brake power.

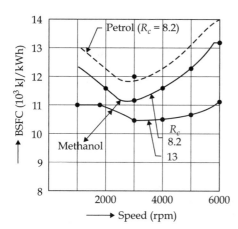

Figure 7.15   Effect of speed on BSFC.

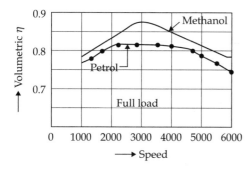

Figure 7.16   Effect of speed on volumetric efficiency.

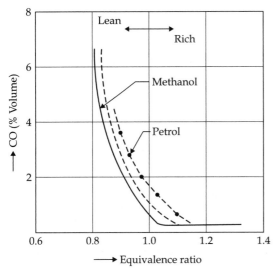

Figure 7.17   Effect of equivalence ratio on CO.

occurs while operating at a lean A:F ratio, whereas in an engine using alcohol, the engine can manage to handle leaner mixtures without any misfire. Important objectionable emissions are CO, HC, NO_x, and aldehydes. The effect of equivalence ratio on all these emissions for petrol and methanol are shown in Figs. 7.17 through 7.20.

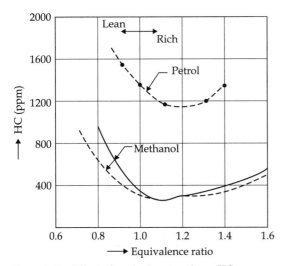

Figure 7.18   Effect of equivalence ratio on HC.

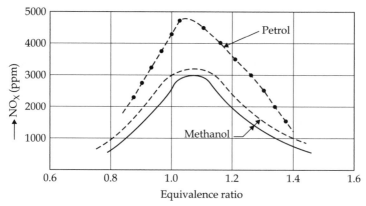

Figure 7.19   Effect of equivalence ratio on NO_x.

For all the above graphs, the engine details and compression ratio are as follows:

1. Full throttle rpm = 2500
2. Compression ratio

Methanol	**Petrol**
$R_c = 9$	$R_c = 9$
$R_c = 12.6$	

Regarding emission, ethanol and methanol are considered as clean fuels, as emissions of CO, HC, and NO_x are reduced by nearly 10–15% compared with a petrol engine. The flame speed of alcohol mixtures is higher than a petrol A:F mixture, and this helps in making the combustion more complete without misfiring.

Regarding the production of formaldehyde, its percentage in exhaust is much higher, which is a great problem to extract methanol in pure form

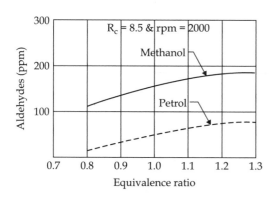

Figure 7.20   Effect of equivalence ratio on aldehyde.

as a replaceable fuel. To avoid this, blends (15–25%) of both alcohols are preferred over pure ethanol or methanol. The properties of blends and their effect lie in between pure alcohol and petrol. As we know, methanol blends have lower stoichiometric air requirements compared to petrol. Therefore, if we use a methanol–petrol blend without any modification in the carburetor, we get more air for combustion, which will reduce the emission of CO and HC as well as NO_x as the engine works cooler with the blend compared with a petrol engine.

Oxidizing catalytic devices can control aldehyde emissions. Platinum–rhodium and platinum–palladium catalysts are considered the most effective in tackling aldehyde emissions of methanol fueling.

Concerning the alcohol fuels, the following conclusions can be drawn:

1. Alcohol is potentially a better fuel than gasoline for SI engines.
2. Its use improves the thermal efficiency as a higher compression ratio (12:16) can be used.
3. It can avoid knocking even at a higher compression ratio because of the high octane number.
4. It provides better fuel economy and less exhaust emissions.
5. High latent heat of alcohol reduces the working temperature of the engine.
6. It gives more power, specially when used as a blend.
7. Easy availability of raw materials.
8. Cost of production is low because of the price hike in crude petroleum.

In agricultural countries like India, we can get ethyl alcohol easily from vegetables, agricultural material, and sugarcane waste at a much lower cost compared with the cost of petrol today. Therefore, replacing petrol with alcohol in a SI engine has a good future.

7.9 Comparison of Ethanol and Methanol

Most of the properties are similar, with differences of only 5–10%. Ethanol is superior to methanol as it has a wider ignition limit (3.5–17) than methanol (2.15–12.8). Its calorific value (CV) (26,880 kJ/kg) is considerably higher than methanol (19,740 kJ/kg).

Ethanol is a much more superior fuel for diesel engines as its cetane number is 8 compared to the cetane number of 3 for methanol. There are wide resources for manufacturing ethanol compared with methanol. Therefore, ethanol is widely used as SI engine fuel in many countries. Methanol is superior to ethanol in one respect: Its vaporization rate is much higher than ethanol. Therefore, mixing with air rapidly forms a uniformly vaporized mixture and also burns uniformly. One major drawback of methanol is that it creates vapor locks because of the higher vaporization rate. Properties of ethanol and methanol as compared with petrol are listed in Table 7.3.

7.10   Ecosystem Impacts Using Alcohol Fuels

7.10.1   Aquatic system impacts

The biological consequences of alcohol spills or leaks into marine water are sensitive to many factors such as scale and duration of the spill, tidal patterns, water currents, flow rate, temperature, and available oxygen. Marine life can tolerate low concentrations of alcohol.

In general, methanol and ethanol are significantly less toxic than gasoline or crude oil. Because alcohols are miscible, volatile, and degradable, they are dispersed readily, and diluted and neutralized in aquatic environments. The aquatic environment recovers more rapidly and completely from an alcohol spill than from a gasoline or crude oil spill of the same volume.

7.10.2   Terrestrial system impacts

The direct exposure of soils to methanol spills results in immediate damage of surface vegetation. The miscibility, volatility, and degradability of alcohols reduce the alcohol residence time in soil and minimizes the environmental impact. Fungal and bacterial populations, which are important agents of nutrient cycling, exhibit 80–90% recovery with 3 weeks of exposure. Total recovery of the site occurs within a period of weeks or months. In comparison, recovery of biodegradation by crude oil and petroleum products takes months or years.

7.10.3   Occupational health impacts

Occupational heath risks associated with using alcohol fuels are lower than those associated with conventional fuels. The relative toxicity of alcohol fuels depends on the means of exposure, inhalation, and ingestion. Gasoline poses a greater occupational health risk than either methanol or ethanol as carcinogens in gasoline can be readily absorbed by the skin or inhaled.

7.10.4   Occupational safety impacts

Two major safety hazards of all fuels are fire and explosion, which can occur because of improper fuel storage, spills, or vehicle accidents. The properties of alcohols and gasoline that pertain to fire and explosion risks include the flash point, auto-ignition temperature, flammability limits, and saturated vapor concentrations. While ethanol and methanol have broader flammability limits than gasoline, gasoline poses a greater risk of fire in open air. Because of the low flash point and auto-ignition temperature of gasoline, gasoline is more likely to ignite and burn rapidly; therefore, the fire hazard is greater for gasoline.

Alcohol-fueled fire can be more readily contained than a gasoline-fueled fire of equivalent volume because alcohols have a lower heat of combustion than gasoline and less of the energy released is converted to radiant heat. Therefore, energy release and potential damage from an explosion caused by alcohol would be less than that of an explosion caused by gasoline.

7.10.5 Socioeconomic impacts

Substitution of alcohol fuels for conventional fuels will increase the number of jobs in fuel production, distribution, and handling industries. Alcohol fuels are expected to cost more than gasoline over the next 10 years.

As a result, vehicle-operating costs will be somewhat higher if alcohol blends are used. The price of alcohol blends varies significantly, depending upon the type of alcohol and feedstock used. Blends containing methanol derived from coal are the least expensive. The most expensive are alcohol blends containing ethanol produced from corn.

7.10.6 Transportation and infrastructure impacts

The existing fuel distribution system must be modified and expanded to accommodate the increasing use of alcohol fuels in the long run. The changes required will include construction of new pipelines, storage facilities, and retrofitting of existing facilities with alcohol-compatible pumps, hoses, valves, and other components.

The vehicle support services such as refueling, maintenance, repairs, and vehicle sales will be unaffected by the use of alcohol fuels. The use of alcohol fuels is not expected to have a significant impact on the existing transportation system infrastructure.

References

1. A. Nag. *Analytical Techniques in Agriculture, Biotechnology and Environmental Engineering,* New Delhi, India: Prentice-Hall of India, 2006.
2. A. Nag. *Text book* of *Agriculture Biotechnology,* New Delhi, India: Prentice-Hall of India, 2007 (in press).
3. E. S. Lipinsky. Chemicals from biomass: Petrochemical substitution options, *Science* **212**, 1465–1471, 1992.
4. B. B. Ghosh, E. V. Thomas, and S. Natarajan. The Performance of a Tractor Diesel Engine with Ethanol Fumigation, Ph.D. Thesis, Mechanical Engineering Department, Indian Institute of Technology, Kharagpur, India, 1992.

Cracking of Lipids for Fuels and Chemicals

Ernst A. Stadlbauer and Sebastian Bojanowski

8.1   Introduction

Lipids [1] in the form of fat and edible oils are important energy sources for humans due to the high calorific value of triacylglycerols (~37 kJ/g, or ~9 kcal/g) and the nutritional benefits of both essential fatty acids and phosphate. In addition, energy stored in lipids may be technically realized by either direct use in combustion or by upgrading into a more versatile fuel. In this respect, lipids play an important role for providing lighting and warmth.

Historically, whale oil lamps and tallow candles were gradually displaced by kerosene lamps and electric bulbs [2]. Nowadays, lipids are attracting interest as a renewable source of fuels and chemical feedstock. Therefore, segmentation in the marketplace for lipids is noticeable [3]. In emerging economies of eastern Asia, there is a demand for cheap, edible commodity oils, such as soybean or palm oils. In developed economies, a nutritionally led demand for niche oils, such as low-trans-fat oils, high-omega-3 oils, and enhanced lipophilic vitamins (especially A and E), prevails. More recently, nonedible uses of lipids arise from the proliferating demand for alternative fuels [4] to substitute liquid hydrocarbons derived from mineral oil [5]. Such strategies fall into four broad categories. One is aimed at fueling diesel engines with pure vegetable oils [6] or vegetable oil–fossil fuel blends [7]. The other focuses on biodiesel (alkyl esters of fatty acids), which is mainly sourced from rapeseed and palm oils [8–10]. Problems [11] associated with the more polar characteristics of vegetable oil and biodiesel in comparison to conventional diesel has given rise to studies for cracking of lipids (vegetable oils/animal fat) into nonpolar hydrocarbons [12] to be used as a base for fuels or chemical commodities. Decomposition studies with and without catalysts (metallic salts, metal oxides) have been performed. Finally, lipids (and proteins) in dead cellular matter such as sewage sludge or meat and bonemeal may be converted by natural catalysts present in the substrate to oil having properties similar to diesel fuel [13].

In the following sections, basic processes of converting lipids into nonpolar hydrocarbons with alkanes, alkenes, and arenes as main constituents are discussed. Details of pure vegetable oils or biodiesel are outlined elsewhere (see Chaps. 4, 5, 6).

8.2 Thermal Degradation Process

Thermal decomposition of vegetable oil was performed to prove the theory of the origin of mineral oil from organic matter [14] as early as 1888. Literature up to 1983 has been reviewed by Schwab et al. [15]. In many cases, inadequate characterization of products formed in pyrolysis of vegetable oils was found. Therefore, analytical data obtained by gas chromatography–mass spectrometry (GC-MS) from thermally decomposed soybean oil and high oleic safflower oil in the presence of air or nitrogen were reported [15].

The ASTM standard method for distillation of petroleum products D86-82 has been used for decomposition experiments. Catalytic systems were excluded in this destructive distillation. The actual temperature of the oil in the feeder flask was about 100°C higher than the vapor temperature throughout the distillation. Under these conditions, GC-MS analysis showed that approximately 75% of the products were made up of alkanes, alkenes, aromatics, and carboxylic acids with carbon numbers ranging from 4 to more than 20 (see Table 8.1).

A comparison of fuel properties is given in Table 8.2. The carbon-hydrogen ratio shows 79% C and 11.88% H for the pyrolyzate of soybean

TABLE 8.1   Composition Data of Pyrolyzed Oil

Class of compounds	Percent by mass high oleic safflower		Soy	
	N_2 sparge	Air	N_2 sparge	Air
Alkanes	37.5	40.9	31.3	29.9
Alkenes	22.2	22.0	28.3	24.9
Alkadienes	8.1	13.0	9.4	10.9
Aromatics	2.3	2.2	2.3	1.9
Unresolved unsaturates	9.7	10.1	5.5	5.1
Carboxylic acids	11.5	16.1	12.2	9.5
Unidentified	8.7	12.7	10.9	12.6

TABLE 8.2   Comparison of Fuel Properties

ASTM test no.	Specification	Distilled soybean oil (N_2 sparge)	No. 2 diesel fuel	Soybean oil	High oleic safflower oil
D613	Cetane rating	43[*]	40 (min.)	37.9[*]	49.1
	Higher heating value, BTU/lb	17,333	19,572	17,035	17,030
D129	Sulfur, %	<0.005	<0.5	0.01	0.02
D130	Copper corrosion, 3 h at 50°C standard strip	1[*]	<3	1[*]	1[*]
D524	Carbon residue at 10% residium	0.45%	<0.35%	0.27%	0.24
D1796	Water and sediment, % by volume	0.05	<0.05	Trace	Trace
D482	Ash, % by weight	0.015	<0.01	<0.01	<0.01
D97	Pour point, °C	+7	−7C (max.)	+12	−21
D445	Viscosity, mm^2/s at 38°C	10.21	1.9–4.1	32.6	38.2
DE191	Carbon, %	79.00	86.61	—	—
	Hydrogen, %	11.88	13.20	—	—

[*]ASTM test D613 with ignition delays observed visually.

oil. This indicates considerable amounts of oxygenated compounds in the distillate. Consequently, methylation of these oils has revealed 9.6–12.2% of carboxylic acids ranging from C-3 to over C-18. This is reflected in the higher viscosity compared to diesel.

Mass-spectral fingerprints of the entire pyrolysis product slate from tripalmitin, different vegetable oils, and extracted oils from microalgae confirm that the decomposition of ester bonds in the absence of external catalysts is extensive [16–18]. However, a great variability in primary pyrolysis/vaporization product slates was observed [18].

Thermodynamic calculation in the degradation process shows that the cleavage of C-O bond takes place at 288°C and fatty acids are the main product [19]. The actual pyrolysis temperature should be higher than 400°C to obtain maximum diesel yield [20]. The mechanism of pyrolysis of vegetable oil has been discussed by various authors [9, 15, 19]. Generally, thermal decomposition proceeds through either a free-radical or carbonium ion mechanism. The primary R-COO splits off carbon dioxide. The alkyl radicals (R), upon disproportionation and elimination of ethene, give rise to alkanes and alkenes. The formation of aromatics is facilitated by a Diels-Alder addition of ethene to a conjugated diene formed in the pyrolysis reactions. However, the product mix and product quality are influenced by many factors such as feed pretreatment, heating rate, and temperature. As vegetable oils may contain trace elements, catalytic effects cannot be completely excluded from any thermal degradation process [21].

8.2.1 Catalytic cracking (CC)

In 1979, a paper [22] from the petrochemical industry reported for the first time that high-molecular-weight triglycerides such as corn oil ($C_{57}H_{104}O_6$) and castor oil ($C_{57}H_{104}O_9$) were convertible to a high-grade gasoline when passed over H-ZSM-5, a catalyst. The latter is a synthetic, medium-pore, shape-selective acid catalyst. Lipids were fed with a piston displacement pump at a rate of 2 mL/h with flowing hydrogen (300 mL/h) over 2 mL of H-ZSM-5 catalyst (0.77 g, 14–30 mesh) contained in a vertical Pyrex reactor at atmospheric pressure and $T = 400–450°C$. Paraffins, olefins, aromatics, and nonaromatics could be detected in the product mixture. The distribution of hydrocarbons is similar to selective conversion of methanol into hydrocarbon units with up to 10 carbon atoms per molecule. In all cases, a high degree of BTX aromatics (benzene, toluene, and xylene) was achieved. The precondition for the catalytic conversion is that the molecule penetrate the cavities of microporous zeolite.

This new catalytic approach has paved the way for a variety of applications. A schematic diagram of experimental arrangements for pyrolysis and catalytic conversion is given in Fig. 8.1.

Conversion of different kinds of vegetable oils over medium-pore H-ZSM-5 have been investigated in detail [23–26]. Catalytic cracking of by-products from palm oil mills with a selectivity of 51wt.% toward aromatic hydrocarbon formation has been reported [27]. To achieve higher yields, this type of work was extended to pyrolysis and zeolite conversion of both whole algae and their major components as well as whole seeds and selected vegetable oils [18, 28–31]. Hot vapors from solid organic material (microalgae, seeds, etc.) or vaporized vegetable oils were passed directly over the H-ZSM-5 catalyst. Products of different

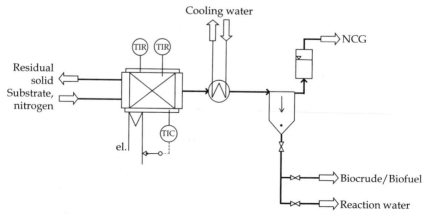

Figure 8.1 General scheme of pyrolysis and catalytic conversion reactor.

algae, seeds, or vegetable oils emerging from the passage showed a uniform, high-octane, aromatic gasoline product. Obviously, the molecular pattern of products is insensitive to the nature of lipids used. This is in contrast to pyrolysis without a catalyst [18].

Upgrading of crude tall oil to fuels and chemicals has been studied at atmospheric pressure and in the temperature range of 370–440°C, in a fixed-bed microreactor containing H-ZSM-5 [32]. The oil was co-fed with diluents such as tetralin, methanol, and steam. High oil conversions, in the range of 80–90 wt.%, were obtained using tetralin and methanol as diluents. Conversions under steam were reduced to 36–70 wt.%. The maximum concentration of gasoline-range aromatic hydrocarbons was 52–57 wt.% with tetralin and steam, but only 39% with methanol. The amount of gas product in most runs was 1–4 wt.% [32].

8.3   Vegetable Oil Fuels/Hydrocarbon Blends

At first glance vegetable oil offers a favorable CO_2 balance. However, when the extra N_2O emission from biofuel production is calculated in "CO_2-equivalent" global warming terms, and compared with the quasi-cooling effect of "saving" emissions of fossil fuel derived CO_2, the outcome is that production of commonly used biofuels can contribute as much or more to global warming by N_2O emissions than cooling by fossil fuel savings [33]. In addition, widespread use of vegetable oil fuels is limited by high viscosity, low volatility, poor cold flow behavior, and lack of oxidation stability during storage [6, 7]. Partial conversion of vegetable oil to hydrocarbons offers the possibility to preserve the favorable environmental characteristics of vegetable oil-based fuels while improving viscosity and cold flow behavior [34, 35]. Figure 8.2 depicts thermogravimetry of vegetable oil without pure oil (dashed line) and in the presence of a Y-zeolite (Koestrolith). The dotted line represents the first derivative from the catalyzed conversion reaction.

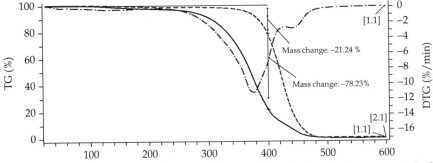

Figure 8.2   Thermogravimetry of commercial vegetable oil fuel without pure oil (dashed line) and in the presence of a Y-zeolite.

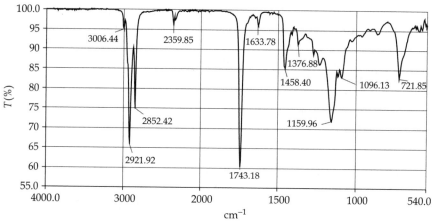

Figure 8.3 IR spectrum of commercial vegetable oil fuel. The bands observed at around 2900 and 1740 cm^{-1} are due to absorption of IR radiation, the absorbed energy causing transitions between energy levels for the stretching vibrations of C-H (hydrocarbons) and C=O bonds (ester function R$_1$COOR$_2$), respectively.

The efficiency of the decarboxylation effect of Y-zeolite activity on pure vegetable oil at $T = 450°C$ may be seen by comparing the IR spectrum of pure vegetable oil fuel in Fig. 8.3 with the corresponding spectrum of the conversion product in Fig. 8.4. The carbonyl band at around 1700 cm^{-1} is an indicator for conversion efficiency.

Table 8.3 summarizes physical and chemical parameters of vegetable oil fuel and conversion products at different temperatures. The change

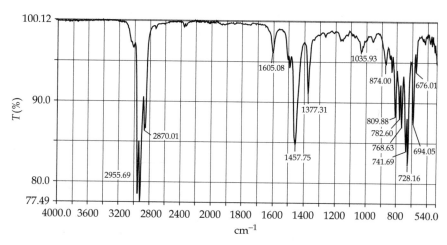

Figure 8.4 IR spectrum of conversion product. The carbonyl function at 1720–1740 cm^{-1} is missing. Absorption bands between 1300 and 650 cm^{-1} are generally associated with complex vibrational and rotational energy changes (fingerprint region) of the molecules.

TABLE 8.3 **Characteristics of Commercial Vegetable Oil Fuel and Its Y-Zeolite Conversion Product**

Parameter	Commercial vegetable oil fuel	Y-zeolite (koestrolith), $T = 430°C$	Y-zeolite (koestrolith), $T = 450°C$
Yield, %	—	34.3	43.5
NCV, MJ/kg	37.0	42.4	42.2
Density, g/mL	0.91	0.79	0.81
Viscosity, mm^2/s	32.87	0.73	0.79
C, %	77.04	87.8	88.32
H, %	12.0	9.59	9.67
N, %	0.29	<0.14	<0.14
S, %	<0.34	<0.34	<0.34

in viscosity is quite remarkable. In accordance with Fig. 8.2, a reaction temperature of $T = 450°C$ is preferred.

8.3.1 Refitting engines

Presently, vegetable oil is regarded as a niche application. One liter of rapeseed oil substitutes for approximately 0.96 L of diesel. The annual yield is 1480 L/ha. CO_2 reduction in relation to the diesel equivalent is about 80% [36]. However, this is questioned in newer literature [33] in terms of global warming reduction considering the effects of extra N_2O entering the atmosphere as a result of using nitrogen-based fertilizers to produce crops for biofuels. Before unmodified vegetable oil is used as a fuel, the engine must be refitted for the fuel to correspond to the viscosity and combustion properties of vegetable oil. Refitting concepts include preheating either the fuel and the injection system or the equipment with a two-tank system. The engine is started with diesel and changes to vegetable oil only when the operating temperature has been reached. Blends of pure vegetable oils and a conversion product together with additives (antioxidants) increase oxidation stability, reduce viscosity, and give a better perspective for vegetable oil fuel markets.

8.3.2 Tailored conversion products

The chemical nature of conversion products depends both on the structure or type of the zeolite used and the reaction temperatures, because restructuring occurs at the inner surface, which acts as a reaction vessel at the molecular scale. Specific reactions depend on the diameters of pores, the resident time of molecules within the pores or channels and voids of the microporous zeolite, and the temperature. The penetration of lipids into a zeolite is depicted in Fig. 8.5. The scheme is based on [22].

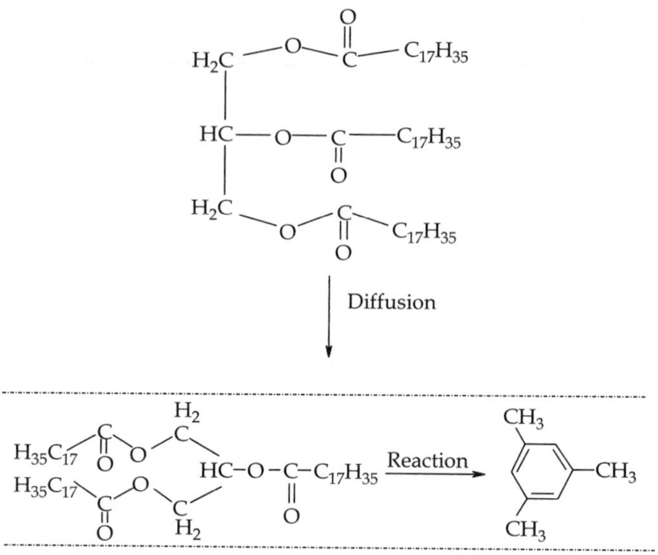

Figure 8.5 Scheme of restructuring triglycerides with shape-selective H-ZSM-5 to aromatic hydrocarbons.

To demonstrate this influence of catalysts and reaction temperature on yields and products, Table 8.4 considers a shape-selective zeolite type H-ZSM-5, commercially available as Pentasil, PZ-2/50H, and Y-zeolite (DAY-Wessalith). The physical characteristics of oils formed from the conversion of animal fat (rendering plant) are depicted [56]. Yields are between 30% and 70%, depending on the type of zeolite and temperature. Net calorific values are in the range of 40 MJ/kg compared to

TABLE 8.4 Yields and Physical Characteristics of Hydrocarbons from Catalytic Conversion of Animal Fat Using Zeolite Types H-ZSM-5 (Pentasil, PZ-2/50H) and DAY-Wessalith at Different Temperatures

Parameter	H-ZSM-5 PZ-2/50H, $T = 550°C$	H-ZSM-5 PZ-2/50H, $T = 400°C$	DAY-Wessalith, $T = 400°C$
Yield, %	31.48	56.74	72.9
NCV, MJ/kg	40.1	40	41.3
Density, g/mL	0.83	0.85	0.81
Viscosity, mm^2/s	0.92	1.01	2.29
C, %	88.6	84.5	83.4
H, %	10.7	12.5	13.5
N, %	<0.14	<0.14	<0.14
S, %	<0.34	<0.34	<0.34

35 MJ/kg of animal fat. All reaction products show relatively low viscosity and densities.

Products at *T* = 400°C.   Again, the chemical nature of products formed from animal fat was analyzed by spectroscopic methods (see Fig. 8.6). The IR spectrum reveals the hydrocarbon nature of products. The strong C-H stretching vibrations (frequencies) at 2900 cm^{-1} is characteristic for alkanes. Functional groups are widely missing. The comparison to diesel from a commercial gas filling station (imprinted spectrum) shows a similar pattern [37].

Proton resonance spectroscopy depicts the chemical environment of protons in the product formed from the conversion of animal fat. Figure 8.7 shows the dominance of aliphatic protons at chemical shifts of 0.9–2.25 ppm. Aromatic protons absorb at 6.5–8 ppm. The inspection of the ratio of the integral of absorptions reveals 5% aromatics for catalysis at *T* = 450°C. This is also reflected in the ^{13}C-NMR spectrogram (see Fig. 8.8). However, with increasing temperature in the catalytic bed, the content aromatic alkylbenzenes increase.

Using ^{13}C-NMR spectroscopy in-depth mode (see Fig. 8.9), negative signals at 30–20 ppm are characteristic for CH$_2$-groups. The intensity indicates the presence of long-chain hydrocarbons. Peaks between 140 and 120 ppm denote carbon atoms of aromatic systems. The low intensity reflects the low content. Obviously, catalytic cracking over a Y-zeolite widely preserves hydrocarbon moiety in vegetable oil.

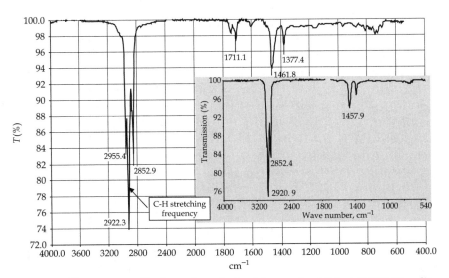

Figure 8.6   IR spectrum of hydrocarbons derived from animal fat at 400°C (Y-zeolite catalyst, DAY-Wessalith).

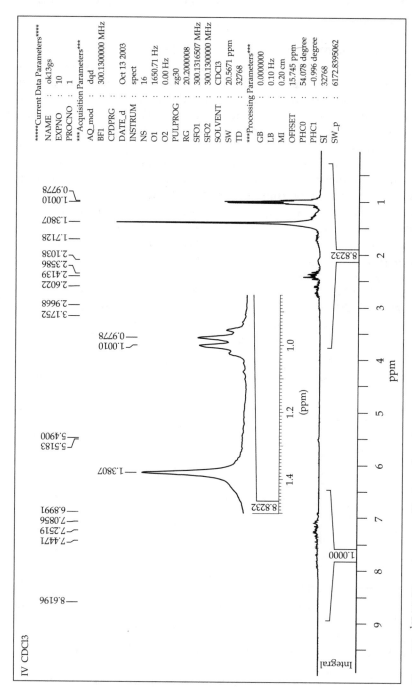

Figure 8.7 ¹H-NMR spectrogram of hydrocarbons from animal fat at $T = 400°C$ (Y-zeolite catalyst, DAY-Wessalith).

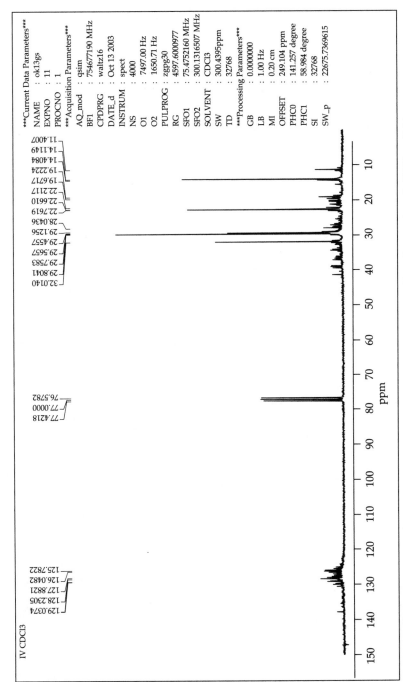

Figure 8.8 ^{13}C-NMR spectrogram of hydrocarbons from animal fat at $T= 400°C$ (Y-zeolite catalyst, DAY-Wessalith).

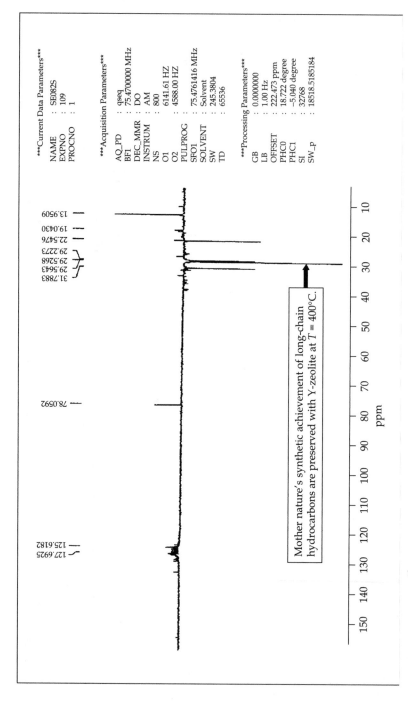

Current Data Parameters

NAME	:	SE0825
EXPNO	:	109
PROCNO	:	1

Acquisition Parameters

AQ_PD	:	qseq
BF1	:	75.4700000 MHz
DEC_MMR	:	DO
INSTRUM	:	AM
NS	:	800
O1	:	6141.61 HZ
O2	:	4588.00 HZ
PULPROG	:	
SFO1	:	75.4761416 MHz
SOLVENT	:	Solvent
SW	:	245.3804
TD	:	65536

Processing Parameters

GB	:	0.0000000
LB	:	1.00 Hz
OFFSET	:	222.473 ppm
PHC0	:	18.722 degree
PHC1	:	−5.040 degree
SI	:	32768
SW_p	:	18518.5185184

13.9509
19.0430
22.5476
29.2273
29.5268
29.5643
31.7883

78.0592

125.6182
127.6925

Mother nature's synthetic achievement of long-chain hydrocarbons are preserved with Y-zeolite at $T = 400°C$.

Figure 8.9 DEPT-135 ^{13}C-NMR spectrogram of biofuel from animal fat at 400°C (Y-zeolite catalyst, DAY-Wessalith).

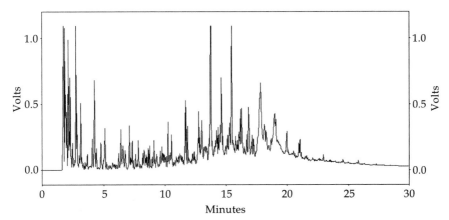

Figure 8.10   GC pattern of Y-zeolite conversion product of animal fat at reaction temperature $T = 400°C$. GC-14A Shimadzu, column: FS-Supreme-5/H53, 30 m; temperature program: 50°C (5 min); 15°C/min to 320°C (10 min); FID detector at 320°C.

These spectroscopic findings are confirmed by gas chromatography (GC) [56]. Pyrolyzates (see Fig. 8.10) and commercial diesel (see Fig. 8.11) have a similar GC pattern. However, crude conversion products contain more volatile hydrocarbons.

GC separation on an OV101 capillary [column: 20 m × 0.3 mm, split 1:25; temperature program: 25°C (2 min), 4°C/min to 320°C] reveals double peaks in more detail (see Fig. 8.12). The first peak is for the alkene with a double bond of a given C number. The second peak is for the alkane having the same C number.

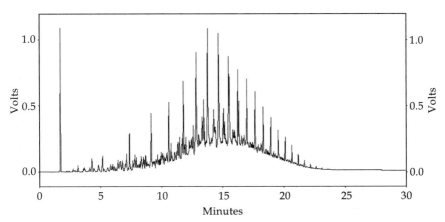

Figure 8.11   GC pattern of commercial diesel. GC-14A Shimadzu, column: FS-Supreme-5/H53, 30 m; temperature program: 50°C (5 min); 15°C/min to 320°C (10 min); FID detector at 320°C.

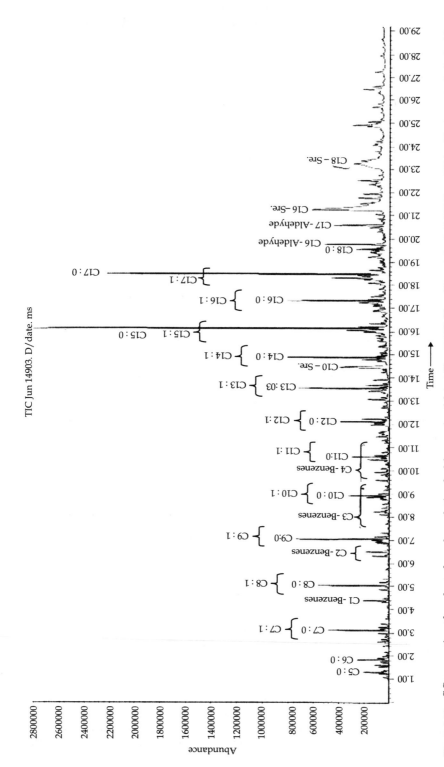

Figure 8.12 GC separation of products from degradation fuel from animal fat at $T = 400°C$ (Y-zeolite catalyst, DAY-Wessalith) with characteristic double peaks.

You may use these hydrocarbons as a base for biofuels. However, there are markets for certain fractions of this hydrocarbon mixture. For example, the C-12 to C-18 fraction is a raw material widely used for bulk commodities. As mineral oil prices increase, it is becoming more financially viable to produce chemical feedstock for commodities and specialities from wastes. Wastes are an energy and carbon source of the future.

Products at $T = 550°C$.   For a given H-ZSM-5-zeolite, the nature of conversion products of lipids (animal fat) shifts to more aromatic compounds as the temperature increases. This is demonstrated by different NMR findings [56] for animal fat as a substrate at a reaction temperature of $T = 550°C$ (see Figs. 8.13 through 8.15). Especially, DEPT-135 ^{13}C-NMR pattern of oil from catalytic conversion of animal fat at 550°C shows the dominance of aromatic protons and a very low amount of CH_2 groups. Chromatographic separation revealed alkylbenzenes (especially 1,3,5-trimethybenzene) as main products [38].

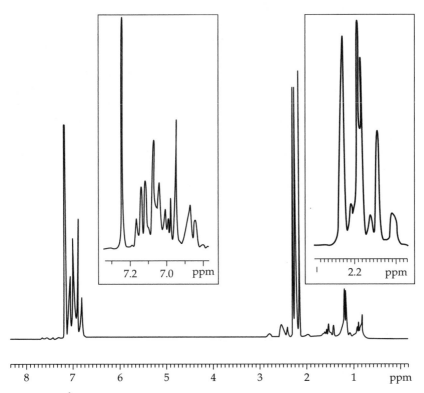

Figure 8.13 ^{1}H-NMR spectrogram of hydrocarbons from animal fat at $T = 550°C$ with the commercial catalyst H-ZSM-5 (Pentasil, PZ-2/50H).

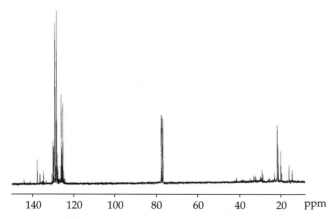

Figure 8.14 ^{13}C-NMR spectrogram of hydrocarbons from animal fat at $T = 550°C$ with the commercial catalyst H-ZSM-5 (Pentasil, PZ-2/50H).

Heating oil and a conversion product from animal fat have been used in a commercial burner (Buderus, Germany). Both oils resulted in emissions within legal limits (see Table 8.5).

A straightforward approach to apply vegetable oil in the most-talked-about biomass-to-liquid-fuel scheme is to use it as a co-substrate in mineral oil refineries. Advantages are low investments for peripheral facilities such as loading and storage and use of an existing infrastructure for distribution and marketing. The processing of rapeseed oil as a feed component in a hydrocracker was described in 1990 [39]. The results are summarized in Table 8.6.

It is worth mentioning that rapeseed oil is converted in the hydrotreatment step to paraffins. The oxygen content of the vegetable oil causes an increased consumption of hydrogen to form water. Changes in quality

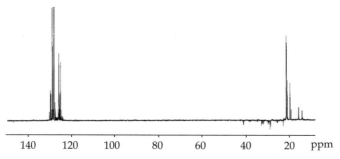

Figure 8.15   DEPT-135 ^{13}C-NMR spectrogram of hydrocarbons from animal fat at $T = 550°C$ with the commercial catalyst H-ZSM-5 (Pentasil, PZ-2/50H).

TABLE 8.5   Comparison of Combustion Parameters: Heating Oil versus Oil Derived from Y-Catalytic Conversion of Animal Fat (AF) at T = 400°C

Parameter	Heating oil	Heating oil & oil from AF 1:1	Oil derived from AF	Limiting value
NCV, MJ/kg	42.0	41.5	41.3	>42[*]
Kinetic viscosity, mm^2/s	3.25	2.74	2.51	<6.0[*]
C, %	86.5	85.7	83.4	—
H, %,	14.0	13.8	13.5	—
N, %	<0.14 (d/l)	<0.14 (d/l)	<0.14 (d/l)	—
S, %	<0.34 (d/l)	<0.34 (d/l)	<0.34 (d/l)	<0.20[*]
NO$_X$, mg/m^3	162	186	233	250[†]
SO$_2$, mg/m^3	87	26	0	350[†]
Smoke pot no.	0.0	0.4	0.4	1[‡]

[*]DIN 51 603
[†]TA Luft
[‡]1. BImSchV
d/l: detection limit

occur in the middle distillate. A lower density and a higher cetane number are a quality-enhancing advantage. A drawback is the susceptibility to freezing point of the fuel. This kind of cold flow behavior would make its use in winter impossible unless special additives are supplemented [40].

8.3.3   Feed component in FCC

In 1993, the influence of 3–30% rapeseed oil in vacuum distillate FCC feed on product slate and quality both at laboratory and at a continuously operated bench-scale apparatus was reported for the first time [41]. On the one hand, results showed decreasing yields of liquid hydrocarbons with increasing rapeseed oil concentrations. On the other hand,

TABLE 8.6   Product Quality of the Hydrocracker with 20% and without Rapeseed Oil as a Feed Component

Fraction	Total oil		Gasoline		Middle distillate		VGO*	
Rapeseed oil, %	0	20	0	20	0	20	0	20
Density (15°C), g/mL	0.815	0.815	0.753	0.759	0.830	0.817	0.852	0.847
Carbon, mass %	86.04	85.33	85.39	85.31	86.06	85.27		
Hydrogen, mass %	14.01	14.42	14.48	14.64	13.82	14.66		
Sulphur, ppm	284	114	29	39	103	18	38	11
Nitrogen, ppm	<1	2	<1	0.5	<1	<1	0.7	<1
Oxygen, mass %	0.1	0.1	0.05	0.1	<0.1	0.06		
NCV, MJ/kg			43.9	44.0	43.4	44.0		
Octane number (MOZ)			63.2	61.4				
Cetane number					48	64		
Pour point,°C					−35	+3		

*VGO, vacuum gas oil.

the gasoline portion in the liquid product increased. Considering propenes, butanes, and *i*-butenes as gasoline potentials, low rapeseed oil portions in the FCC feed seem to result in an optimum yield of gasoline plus gasoline potentials. Most interestingly, the gasoline fraction recovered from a 500-h bench scale run using a feed with 30% rapeseed oil proved suitable for standardized gasoline blending. Calcium concentration $c(Ca) > 2$ ppm gradually decreases FCC catalyst activity. Oxygen contained in the vegetable oil was mainly converted to water. Moreover, traces of phenols and carboxylic acid were detected in the liquid reaction product.

MAT with animal fat.   In a laboratory scale, mixtures of vacuum gas oil and up to 15% of animal fat were converted in a Micro-Activity Test (MAT) unit [37]. Results are given in Figs. 8.16 and 8.17. Two aspects are of special interest. First, yields of propene and butene increase with animal fat as a co-substrate. This is an advantageous finding as C-3 and C-4 are gasoline potentials. C-3 and C-4 liquefied petroleum gas can be used for the manufacture of isoparaffins for motor gasoline through alkylation and polymerization processes.

Second, a higher yield of gasoline fraction is observed. This is a consequence of the high hydrogen:carbon ratio of about 2 and the low heteroatom content. For this reason, biomaterials with a hydrocarbon-like structure are particularly interesting candidates for conversion to low-molecular-weight fuels or chemical raw materials. Problems to be investigated are possible calcium and phosphate deposits on the catalyst particles which may impair catalyst activity and process stability of the riser. Therefore, the process must include a regeneration step. The market will decide whether or not animal fat can substitute a bit of non-renewable resources in petroleum refining.

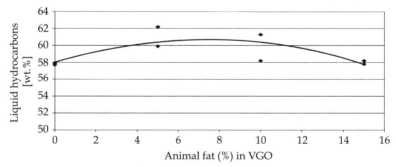

Figure 8.16   Cocatalytic cracking of animal fat and vacuum gas oil (VGO) in MAT experiments. At around 7% feed component, the maximum yield of liquid hydrocarbons is found; weight–hourly space velocity (WHSV) = 2 h^{-1}.

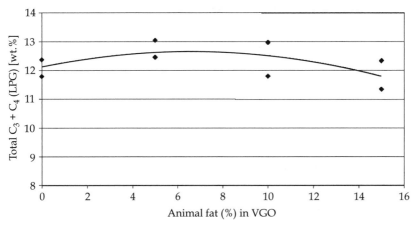

Figure 8.17 Cocatalytic cracking of animal fat and vacuum gas oil (VGO) in MAT experiments. Gasoline potentials show a maximum at low rapeseed portions around 7%; weight–hourly space velocity (WHSV) = 2 h^{-1}.

8.4   Other Metal Oxide Catalysts

Cottonseed oil has been thermally decomposed at 450°C using 1% Na_2CO_3 as a catalyst [42]. Pyrolysis produced a yellowish-brown oil with 70°C yield. The fuel properties of original and pyrolyzed cottonseed oil are summarized in Table 8.7. Results of ASTM distillation compared to diesel are given in Table 8.8 showing a higher volatility for the conversion product.

Rapeseed oil was pyrolyzed in the presence of about 2% calcium oxide up to a temperature of 450°C [43]. An oil was obtained with a heating

TABLE 8.7   Fuel Properties of Original and Pyrolyzed Cottonseed Oil and No. 2 Diesel Fuel

Property	Diesel fuel	Pyrolyzed oil	Original oil
API gravity	35	35	21.5
Specific gravity (at 15.6°C)	0.8504	0.8500	0.9246
Kinetic viscosity (mm/s^2) at 40°C	0.0213	0.0178	0.0357
Cetane index	33	28	20
Flash point, °C	96	53	268
Sulfur content, wt.%	0.04	Nil	0.02
Pour point, °C	0.0	>15	23
Sediment content, wt.%	Nil	0.04	5.0
Calorific value, kJ/g	45.57	45.57	41.80
Water content, vol.%	Nil	2.98	1.20
Ash content, wt.%	Nil	Nil	Nil
Carbon residue, wt.%	0.01	0.16	1.06

TABLE 8.8   Results of ASTM Distillation of No. 2 Diesel
Oil and Pyrolyzed Cottonseed Oil as Volume Percent

Parameter	Temperature °C	
	Diesel oil	Pyrolyzed oil
Distillate, %		
0	63	55
10	105	79
20	174	116
30	192	131
40	200	157
50	210	178
60	235	186
70	245	220
80	250	247
90	255	269
98	260	–
Recovery, %	98	90
Residue, %	1	9
Loss, %	1	1

value of 41.3 MJ/kg, a kinematic viscosity of 5.96 mm^2/s, a cetane number of 53, and a flash point of 80°C. When tested on a diesel engine, the thermal efficiency (η_{th}) and brake specific fuel consumption were improved. The concentration of nitrogen oxide in the exhaust gas was less than diesel. The absence of sulfur in the pyrolytic oil was seen as an advantage to avoid corrosion problems and the emission of polluting sulfur compounds from combustion.

Triolein, canola oil, trilaurin, and coconut oil were pyrolyzed over activated alumina at 450°C and atmospheric pressure [44]. The products were characterized by IR spectrometry and decoupled ^{13}C-NMR spectroscopy. The hydrocarbon mixture contained both alkanes and alkenes. These results are significant for the pyrolysis of lipid fraction in sewage sludge as well as for wastes from food-processing industries [44].

Pyrolysis of rapeseeds, linseeds, and safflowers results in bio-oil containing oxygenated polar components. Hydropyrolysis at medium pressure in the presence of 1% ammonium dioxydithiomolybdenate $(NH_4)_2MoO_2S_2$ can remove two-thirds to nine-tenths of the oxygen present in the seeds to generate bio-oils in yields up to 75% [45]. In addition, extraction with organic solvents including diesel oil gave yields up to 40%.

The potential of liquid fuels from *Mesua ferrea* seed oil [46], *Euphorbia lathyris* [47, 48], and underutilized tropical biomass [49] has been investigated in the search for "energy farms" involving the purposeful cultivation of selected plants to obtain renewable energy sources.

8.5   Cracking by In Situ Catalysts

This method is applicable for cellular biomass containing lipids, e.g., sewage sludge or organic residues from rendering plants. The European Union is looking for new markets for both materials. On the one hand, treatment of municipal and industrial wastewaters generates huge quantities of sludge, which is the unavoidable by-product especially if biological processes are used. Management of this residue poses an urgent problem. The residue contains about 60% of bacterial biomass and up to 40% of inorganic materials such as alumina, silicates, alkaline and alkaline earth elements, phosphates, and varying amounts of heavy metals [56]. On the other hand, returning animal meal (AM) or meat and bone meal (MBM) from the rendering plant into the food cycle is forbidden by law since the BSE crisis [50, 51]. Besides burning, low-temperature conversion (LTC) of these organic materials offers an alternative disposal method [52–54]. LTC is a thermocatalytic process whereby organics react to hydrocarbons as the main product [12].

The conversion of bacterial biomass or organic residues from rendering plants to oil may be formally defined by considering the starting materials and the end products. The principal components of these substrates are proteins and lipids. They make up about 60–80% of this biomass. The average elemental composition of neutral lipids is $C_{50}H_{92}O_6$. An empirical formula for proteins is $(C_{70}H_{135}N_{18}O_{38}S)_x$. From these compounds, nonpolar hydrocarbons of the general elemental composition C_nH_m have to be produced [13, 55].

Obviously, LTC removes the heteroatoms from both principal components. In general, it splits off functional groups from complex biomass. The process operates at moderate temperatures (380–450°C), essential atmospheric pressure, and the exclusion of oxygen. Under these conditions, heteroatoms from organics are removed as ammonia (NH_3), dihydrogensulfide (H_2S), water (H_2O), and carbon dioxide (CO_2). This decomposition scheme may serve as a model for the formation of coal from primarily plant sources. Carbohydrates ($C_6H_{10}O_5)_n$ are the principal components in plants. The elimination of water from carbohydrates produces elemental carbon, according to the following reaction:

$$(C_6H_{10}O_5)_n - 5H_2O \rightarrow C_m$$

Consequently, carbohydrates of bacterial mass will be converted to carbon, mainly in the form of graphite [56, 57]. Therefore, the formation of oil from complex biomass will always be accompanied by the formation of carbon. Figure 8.18 depicts the mechanism for the production of oil from lipids by LTC [58].

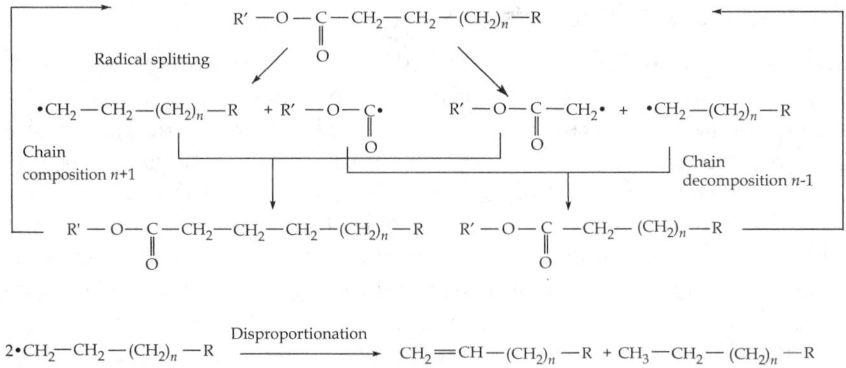

Figure 8.18 Mechanistic aspects of the formation of hydrocarbons by cracking of lipids [58].

It is worth mentioning that the ash content (Table 8.9) includes natural catalysts (e.g., alumina and silicates) that substantially influence the yield and composition of LTC products. Table 8.9 shows results of the conversion of these organic residues. Yields of oil, solid product, water, volatile salts (NH_4Cl, $NaHCO_3$), and noncondensable gases (NCG: CO_2, H_2, C-1–C-4 alkanes and different alkenes) are given in Fig 8-19. Digested sludge produces less oil than aerobically stabilized sludge. This correlates with the carbon content in Table 8.9. The food chain of anaerobic bacteria efficiently removes organic carbons as biogas (CH_4/CO_2). Thus it is no longer available for the production of oil in subsequent LTC. AM shows higher yields of oil due to its higher content of fat and proteins (Table 8.9). The viscosities of untreated oils at 40°C are as follows: DS, 14 mm^2/s; AS, 35 mm^2/s; AM, 27 mm^2/s; and MBM, 21 mm^2/s. In comparison, diesel from a filling station has a viscosity of

TABLE 8.9 Chemical and Physical Characteristic Substrates for LTC

Parameter	AS	DS	AM	MBM
Dry solids, %	95.0	79.6	94.3	95.0
Ash content, %	35.1	40.7	23.2	38.2
Protein, %	32.9	26.6	52.3	49.6
Fat, %	—	—	14.4	8.9
Calcium as Ca, %	—	9.6	9.7	20.1
Phosphorus as P_2O_5, %	—	6.3	8.7	16.0
NCV, MJ/kg	14.2	9.9	18.8	15.4
C, %	31.6	23.0	42.5	30.5
H, %	4.4	5.0	6.6	4.8
N, %	5.0	3.3	8.3	7.6
S, %	0.6	1.0	0.5	0.3

AS: aerobically stabilized sewage sludge; DS: digested sewage sludge; AM: animal meal; MBM: meat and bone meal.

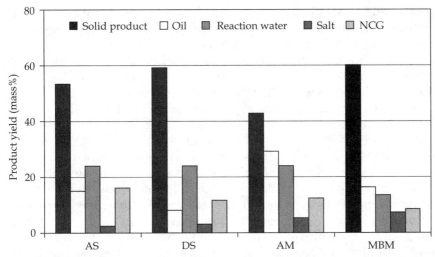

Figure 8.19 Mass yield of LTC products from different substrates. AS: Aerobically stabilized sewage sludge; DS: digested sewage sludge; AM: animal meal; MBM: meat and bone meal [60].

about 4 mm^2/s. The solid products consist of carbon, nonvolatile salts (e.g., CaKPO$_4$), and metal oxides or sulfides. Especially in the case of AM and MBM, the solid product is of commercial interest due to its high content of phosphate. It is free of proteins [59].

As with natural crude oils, the hydrocarbon mixtures obtained by LTC of lipids containing biomass are of a highly complex composition. For example, Fig. 8.20 shows the gas chromatogram of oil derived from

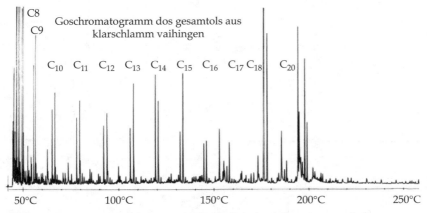

Figure 8.20 Chromatographic separation of oil derived from sewage sludge; separation was performed on OV101 capillary column 20 m × 0.3 mm, split 1:25; temperature program 25°C (2 min), 4°C/min to 320°C.

sewage sludge AS [61]. Peaks assigned by numbers correspond to the aliphatic, unbranched saturated hydrocarbons. The peak appearing before the *n*-alkane corresponds to the *n*-alkenes.

The predominant aliphatic nature of oils produced is readily ascertained by NMR spectroscopy. Figure 8.21 depicts the ¹H-NMR spectrogram of oil from DS with about 5% of aromatic protons.

Infrared spectroscopy (see Fig. 8.22) reveals the presence of C-H-stretching frequencies at 2850–3000 cm^{-1}. In addition, the spectrum provides clear evidence of hydrogen bonding due to a broad absorption band of 3350 cm^{-1}. Thus, decarboxylation of lipids in the presence of in situ catalysts is not complete. This is consistent with the higher viscosities in comparison to diesel. A special loop reactor for recycling catalytic activity to overcome these problems has been designed [62].

Hydrocarbons are derived from both lipids and proteins in the sewage sludge in the presence of in situ catalysts. However, oil produced from proteins under anaerobic LTC conditions is high in nitrogen and sulfur: Amines, purins, and mercaptanes are trace contaminants that are formed. Consequently, this oil smells and is a nuisance, and upgrading (e.g., over H-ZSM-5 as catalyst) is essential [64]. The useful oil is

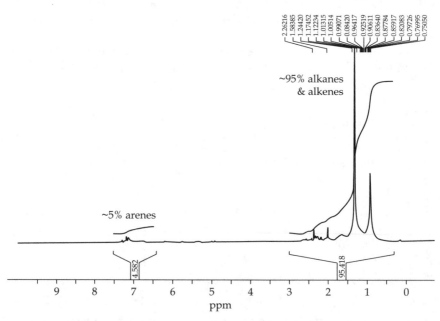

Figure 8.21   ¹H-NMR of oil from LTC of DS at $T = 400°C$.

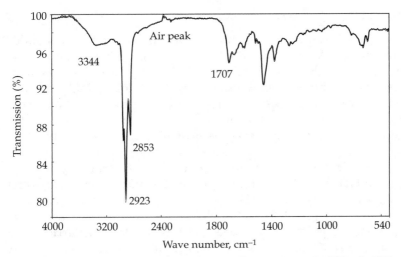

Figure 8.22   Infrared spectrum of oil from DS shows associated –OH and –NH bonds (3350 cm^{-1}) from the remaining carboxylic acids R-COOH or amides R-CONH$_2$ [63].

produced from lipids. When sewage sludge was spiked with triolein, representative of unsaturated triglycerides, the compound did not survive the LTC [65]. As a result, sludge was extracted with toluene using a Soxleth extraction method to yield 12 wt.% lipids. Pyrolysis of sewage sludge lipids over activated alumina produced liquid hydrocarbons containing mostly alkanes [65]. Even the carboxylic acid fractions of the lipids that were separated were completely converted. This is in contrast to direct sewage sludge LTC, where long-chain carboxylic acids are detectable in the IR spectrum (see Fig. 8.22). The reason is the lower content of catalytically active in situ material. Pyrolyzed liquid products from sewage sludge lipids contain virtually no nitrogen or sulfur (see Table 8.10). Only this liquid has a potential for use as a base for commercial fuels [65].

TABLE 8.10   Elemental Composition of Original Dried Sludge, Extracted Lipids, and Pyrolyzed Liquid Product

Component	C, %	H, %	O, %	N, %	S, %	Ash, %
Original sludge	39.5	6.0	26.1	6.00	0.7	20.5
Extracted lipids	72.2	10.7	14.2	0.35	0.85	0.6
Liquid product	86.6	13.5	0.0*	0.08	0.22	0.0*

*By difference.

8.6 Conclusion

The potential offered by lipids for alternative fuel and chemicals is widely recognized. Various sources from plant seeds to animal fat are commercially available. Cracking converts polar esters into nonpolar hydrocarbons. Highly efficient conversion technology should include use of catalysts, e.g., zeolites such as H-ZSM-5 or Y-type representatives. At 380–450°C, alkanes and alkenes are predominantly found in the liquid product. With increasing temperatures up to 550°C, the product spectrum shifts to alkylbenzenes with 1,3,5-trimethylbenzene as the main product. For commercial fuel production based on lipids, assessment of oxidation stability and deposit formation are essential. Influences on regulated and nonregulated emissions have to be analyzed. Attention should be paid both to the NO_x content of exhaust gas and to the particle size distribution with special focus on ultrafine particles. In addition, mutagenic tests for potency of particulate matter extracts are recommended. Finally, it has to be kept in mind, that the replacement of fossil fuels by biofuels may not bring the intended climate cooling due to the accompanying emissions of N_2O from the use of N-fertilizers in crop production. Much more research on the sources of N_2O and the nitrogen circle in connection with biofuels from lipids is needed.

References

1. A. Lehninger, M. M. Cox, M., and O. Lehninger. *Principles of Biochemistry*, New York: W. H. Freeman & Company, 2004.
2. E. S. Lipinsky, D. Anson, J. R. Longanbach, and M. Murphy. Thermochemical applications for fat and oils, *Journal of American Oil Chemists' Society* **62**, 940–942, 1985.
3. D. J. Murphy. Production of healthy vegetable oils, In: *Book of Abstracts, 4th European Federation Lipid Congress: Oils, Fats and Lipids for a Healthier Future*, October 1–4, 2006, University of Madrid, Spain, p. 27.
4. Directive 2003/30/EG of European Parliament and Council from May 8, 2003 for Promotion of the Use of Biofuels or Other Renewable Fuels in the Traffic Sector.
5. J. Krahl, A. Minack, N. Grope, Y. Tuschel, and O. Schroeder. Emissions from biodiesel and vegetable oil, In: *Fuels of the Future 2007*, Conference Papers, Domnik Rutz and Rainer Janson (Eds.) UFOP & BBE, Berlin Germany November 27–28, 2006.
6. G. Vellguth. Performances of Vegetable Oils and Their Monoesters as Fuels for Diesel Engines, SAE Paper No. 831358, Society of Automotive Engineers, Inc. Warrendale, PA, 2000
7. K. R. Kaufman, T. J. German, G. L. Pratt, and J. Derry. Field evaluation of sunflower oil/diesel fuel blends in diesel engines, *Transactions of the ADSE* **29**(1), 2–9, 1986.
8. K. L. Harrington. Chemical and physical properties of vegetable oil esters and their effect on diesel fuel performance, *Biomass* **9**, 1–17, 1986.
9. A. Srivasrava and R. Prasad. Triglycerides-based diesel fuels, *Renewable & Sustainable Energy Reviews* **4**, 111–133, 2000.
10. M. B. Wahid. Biofuel production and market in Malaysia—Current developments and outlook, In: *Fuels of the Future 2007*, Conference Papers, UFOP & BBE, Berlin, Germany, November 27–28, 2006.
11. O. Loest, J. Ullmann, and J. Winter. Investigations on the Addition of FAME to Diesel Fuels, DGMK-Research Report 639, Hamburg, Germany, July 2006.

12. E. A. Stadlbauer, S. Bojanowski, M. S. Hossain, and A. Fiedler. Thermocatalytic production of biofuels from animal fat and solid residues from extraction of rapeseeds, In: *Book of Abstracts, 4th European Federation of Lipids Congress: Oils, Fats and Lipids for a Healthier Future*, October 1–4, 2006, University of Madrid, Spain, p. 115.
13. E. Bayer and M. Kutubuddin. Thermocatalytic conversion of lipid-rich biomass to oleochemicals and fuel, In: *Research in Thermochemical Biomass Conversion*, Bridgwater, A. V. and Kuester, J. I. (Eds.), London: Elsevier Applied Science, 1988, pp 518–30.
14. Engler, E., Ber. **22** [1888], 1816.
15. A. W. Schwab, G. J. Dykstra, E. Selke, S. C. Sorenson, and E. H. Pryde. Diesel fuel from thermal decomposition of soybean oil, *Journal of American Oil Chemists' Society* **65**, 1781–1786, 1988.
16. R. J. Evans and T. A. Milne. Molecular characterization of the pyrolysis of biomass. I. Fundamentals, *Energy and Fuels* **1**, 123–137, 1987.
17. R. J. Evans and T. A. Milne. Molecular characterization of the pyrolysis of biomass. II. Applications, *Energy and Fuels* **1**, 311–319, 1987.
18. R. J. Evans, T. A. Milne, and N. Nagle. Catalytic conversion of microalgae and vegetable oils to premium gasoline, with shape-selective zeolites, *Biomass* **21** 219–232, 1990.
19. C. Zhenyi, J. Xing, and L. Li. Thermodynamics calculation of the pyrolysis of vegetable oils, *Energy Sources* **26**, 849–856, 2004.
20. R. O. Idem, S. P. Katikaneni, and N. N. Bakhshi. Thermal cracking of canola oil: Reaction products in the presence and absence of steam, *Energy and Fuel* **10**, 1150–1162, 1996.
21. Bridgwater, A. V. Catalysis in thermal biomass conversion, *Applied Catalysis A: General* **116**, 5–47, 1994.
22. P. B. Weisz, W. O. Haag, and P. G. Rodewald. Catalytic production of high-grade fuel (gasoline) from biomass compounds by shape-selective catalysis, *Science* **1079**, 57–58, 2005.
23. Y. S. Prasad and N. N. Bakhshi. Effect of pretreatment of HZSM-5 catalyst on its performance in canola oil upgrading, *Applied Catalysis A: General* **18**, 71–85, 1984.
24. Y. S. Prasad and N. N. Bakhshi. Catalytic conversion of canola oil to fuels and chemical feedstock. Part I: Effect of process conditions on the performance of HZSM-5 catalysts, *Canadian Journal of Chemical Engineering* **64**, 278–284, 1986.
25. E. C. Novella, G. O. Escudero, J. A. Alonso, F. P. M. Andres, and P. Canizares. Conversion of vegetable oils and extracts to hydrocarbons, In: *Actas Simp. Iberoam. Catal., 9th*, Vol. 2., Lisbon, Portugal: Soc. Iberoam. Catal., 1984, pp 1623–1624.
26. Y. S. Prasad, H. Yaoliang, and N. N. Bakhshi. Effect of hydrothermal treatment of HZSM-5 catalyst on its performance for the conversion of canola and mustard oil to hydrocarbons, *Industrial & Engineering Chemistry Product Research Review* **25**, 257–267, 1986.
27. J. Graille, L. P. Geneste, A. Guida, and O. Norina. Production of hydrocarbons by catalytic cracking of the by-product of the palm oil mills, *Revue Francaise des Corps Gras* **28**, 421–426, 1981.
28. O. Onay, S. H. Beis, and O. Kockar. Fast pyrolysis of rapeseed in a well-swept fixed bed reactor, *Journal of Analytical and Applied Pyrolysis* **58/59**, 995–1007, 2001.
29. O. Onay and O. M. Kockar. Slow, fast and flash pyrolysis of rapeseed, *Renewable Energy* **28**, 2417–2433, 2003.
30. A. E. Putun, O. M. Kockar, S. Yorgun, H. F. Gercel, J. Andresen, and C. E. Snape. Fixed bed pyrolysis and hydropyrolysis of sunflower bagasse: Product yield and compositions, *Fuel Process Technology* **46**, 49–62, 1996.
31. E. A. Stadlbauer, S. Bojanowski, S. Hossain, S. Stengl, B. Weber, A. Bone, B. Jehle, E. Ruenagel, *Stoffstrommanagement und Energieeffizienz, η[energie]*, **1**, 48–52, 2007.
32. R. K. Sharma and N. N. Bakhshi. Upgrading tall oil to fuels and chemicals over HZSM-5 catalyst using various diluents, *Canadian Journal of Chemical Engineering* **69**, 1082–1086, 1991.
33. P. J. Crutzen, A. R. Mosier, K. A. Smith, and W. Winiwarter. N_2O release from agro-biofuel production negates global warming reduction by replacing fossil fuels, *Atmos. Chem-Phys. Discuss.*, **7**, 11191–11205, 2007 (www.atmos-chem-phys-discuss.net/7/11191/2007/).

34. E. A. Stadlbauer, S. Bojanowski, S. M. Hossein. Unpublished Results.
35. W. Funktion and E. A. Stadlbauer. Verfahren und Vorrichtung zur Herstellung von Kohlenwasserstoffen aus biologischem Fett DE 10 2004 012 583 A1 2005.12.01, Patent Office, Munich, 2004.
36. Fachagentur Nachwachsende Rohstoffe E.V. (FNR). Biofuels, Guelzow, Germany, 2006, p. 14.
37. E. A. Stadlbauer, et al. Herstellung von Kohlenwasserstoffen aus Tierfett durch thermokatalytisches Spalten, *Erdöl Erdgas Kohle* **122**, 64–69, 2006.
38. S. Bojanowski, A. Fiedler, A. Frank, J. Rossmanith, G. Schilling, and E. A. Stadlbauer. Catalytic production of liquid fuels from organic residues of rendering plants. In: *Proceedings of Environmental Science and Technology*, Vol. II, pp. 106–107, American Science Press, New Orleans, 2005.
39. M. Rupp. Verarbeitung von Rapsöl in Mineralölraffinerien, VDI Berichte Nr. 704, Germany, pp. 97–111, 1990.
40. D. Schliephake and C. -M. Hacker. Joint project for the assessment of the agricultural, process and chemical engineering framework for utilization of rapeseed oil and its conversion products as fuels, Project 0310026 A, BMBF/BML, Germany, 1994.
41. K. Bormann, H. Tilgner, and H. -J. Moll. Rapeseed oil as a feed component for the catalytic cracking process, *Erdöl Erdgas Kohle* **109**, 172–176, 1993.
42. F. A. Zaher and A. R. Taman. Thermally decomposed cottonseed oil as a diesel engine fuel, *Energy Sources* **15**, 400–504, 1993.
43. O. A. Megahed, N. M. Abdelmonem, and D. M. Nabil. Thermal cracking of rapeseed oil as alternative fuel, *Energy Sources* **26**, 1033–1042, 2004.
44. D. G. Boocock, S. K. Konar, A. Mackay, P. T. Cheung, and J. Liu. Fuels and chemicals from sewage sludge. 2. The production of alkanes and alkenes by pyrolysis of triglycerides over activated alumina, *Fuel* **71**, 191–197, 1992.
45. O. Onay, A. F. Gaines, O. M. Kockar, M. Adams, T. R. Tyagi, and C. E. Snape. Comparison of the generation of oil by extraction and the hydropyrolysis of biomass, *Fuel* **85**, 382–392, 2006.
46. D. Konwer, S. E. Taylor, B. E. Gordon, J. W. Otvos, and M. Calvin. Liquid fuels from *Mesua ferrea* L. seed oil, *Journal of American Oil Chemists' Society* **66**, 223–226, 1989.
47. E. K. Nemethy, J. W. Otvos, and M. Calvin. Hydrocarbons from *Euphorbia lathyris*, *Pure and Applied Chemistry* **53**, 1101–1108, 1981.
48. M. Oberdörfer. Niedertemperaturkonvertierung von Ölsaaten, Doctoral Thesis, Faculty for Chemistry and Pharmacy, University of Tuebingen, Germany, 1990.
49. K. Esuoso, H. Lutz, M. Kutubuddin, and V. Bayer. Chemical composition and potential of some underutilized tropical biomass. I: Fluted pumpkin (*Telfairia occidentalis*), *Food Chemistry* **61**, 487–492, 1998.
50. Amtsblatt der Europäischen Gemeinschaften. Über Maßnahmen zum Schutz gegen die transmissiblen spongiformen Enzephalopathien bei der Verarbeitung bestimmter tierischer Abfälle und zur Änderung der Entscheidung 97/735/EG der Kommission, 1999.
51. E. A. Stadlbauer, S. Bojanowski, A. Frank, S. Skrypsky-Mäntele, and C. Zettel. Treatment of bovine carcasses from veterinary clinics in times of BSE, In: *1st International Symposium on Residue Management in Universities*, November 6–8, 2002, Santa Maria, Brasilien, pp. 31–32.
52. E. A. Stadlbauer, S. Bojanowski, A. Fiedler, A. Frank, and J. Rossmanith. Niedertemperaturkonvertierung von Klärschlamm zu Kohlenwasserstoffen (Treibstoff für Motoren). Fachtagung: Wasser- und Abwassertechnologie—Neue Märkte erschließen, Justus-Liebig-University of Giessen, Germany, IFZ, March 17, 2004.
53. S. Bojanowski. Untersuchungen zur Niedertemperaturkonvertierung von Tiermehl, Diploma Thesis, University of Applied Sciences Giessen-Friedberg, Giessen, Germany, August 2002.
54. J. Piskorz, D. S. Scott, and I. B. Westerberg. Flash pyrolysis of sewage sludge, *Industrial & Engineering Chemistry Process Design and Development* **25**, 265–270, 1986.
55. S. Skrypski-Mäntele. Untersuchung über die Eigenschaften der Konvertierungskohle der Niedertemperaturkonvertierung, Doctoral Thesis, University of Tuebingen, Germany, 1992.

56. E. A. Stadlbauer. Thermokatalytische Niedertemperaturkonvertierung (NTK) von tierischer und mikrobieller Biomasse unter Gewinnung von Wertstoffen und Energieträgern im Pilotmaßstab, Interim Report DBU, Germany Grant No. 18153, Giessen, Germany, November 2003.

57. E. A. Stadlbauer. Thermokatalytische Niedertemperaturkonvertierung (NTK) von tierischer und mikrobieller Biomasse unter Gewinnung von Wertstoffen und Energieträgern im Pilotmaßstab, Final Report DBU, Germany Grant No. 18153, Giessen, Germany, July 2005.

58. L. Rupp. Mechanismus des Thermischen Abbaus von Fetten und Fettsäuren, Doctoral Thesis, University of Tuebingen, Germany, 1986.

59. M. S. Hossain. Neuwertschöpfung durch Treibstoffproduktion aus tierischen Substraten unter Berücksichtigung arbeitshygienischer Aspekte, Diploma Thesis, University of Applied Sciences, Giessen-Friedberg, Giessen, Germany, July 2005.

60. E. A. Stadlbauer, S. Bojanowski, A. Frank, R. Lausmann, and W. Grimmel. Untersuchungen zur thermokatalytischen Umwandlung von Klärschlamm und Tiermehl. KA – Abwasser, Abfall 50, 1558–1562, 2003.

61. M. Kutubuddin. Niedertemperaturkonvertierung von Biomasse zu Öl und Kohle, Doctoral Thesis, University of Tuebingen, Germany, 1982.

62. W. Funktion and E. A. Stadlbauer and Catalytic Reactor, PCT DE 2004/000329, WO 2004/074181 A2, 2004.

63. E. A. Stadlbauer, S. Bojanowski, A. Fiedler, J. Hoogveldt, J. Rossmanith, G. Schilling, A. Frank, and R. Lausmann. Thermocatalytic reactor to convert organic residues from wastewater treatment and rendering plants to bio-fuels and chemicals, In: Proceedings: Anaerobic Digestion 2004, 10th World Congress, August 29–September 3, 2004, Montreal, Canada, pp. 914–920.

64. R. K. Sharma and N. N. Bahkashi. Catalytic upgrading of pyrolysis oil, Energy and Fuel 7, 306–314, 1993.

65. S. K. Konar, D. G. Boocock, V. Mao, and J. Liu. Fuels and chemicals from sewage sludge: 3. Hydrocarbon liquids from the catalytic pyrolysis of sewage sludge lipids over activated alumina, Fuel 73, 642–646, 1994.

Fuel Cells

A. K. Sinha

9.1 Introduction

Global primary energy consumption (i.e., energy used for space heating, transportation, generating electricity, etc.) is expected to triple from about 400 exajoules (EJ = 10^{18} joules) per year in 2000 to about 1200 EJ/yr in 2050 at the present rate of increase in consumption. However, due to increased energy efficiency of the devices, the actual increase is expected to be about 800–1000 EJ.

More than 80% of the present primary energy requirements are met by fossil fuels. The consequences of burning hydrocarbons at such a large scale for our energy needs are already evident in the form of global warming and its disastrous environmental effects. In order to permit stabilization of anthropogenic greenhouse gases, fossil fuel consumption will have to be limited to about 300 EJ/yr by 2050. Hopefully, the concern about global warming, limit on fossil fuel supplies, and rise in their prices will force us to gradually decrease the use of fossil fuels in the future. Reducing hydrocarbon consumption to 300 EJ requires carbon-free energy sources to supply the difference ~700 EJ/yr. This shortfall is a problem that requires immediate attention and proactive action for sustainable development.

The need for an efficient, nonpolluting energy source for transportation, large-scale generation, and portable devices has spurred the development of alternative energy sources. Fuel cells are a promising alternative energy source that fits the above requirements [1–6]. A fuel cell is an electrochemical device that converts the chemical energy of a fuel (hydrogen, natural gas, methanol, gasoline, etc.) and an oxidant (air or oxygen) into electricity, with water and heat as by-products. Since no combustion is involved in the hydrogen fuel cell process, no NO_x are generated. Since sulfur is a poison to fuel cells, it has to be removed from fuel before feeding it to a fuel cell; therefore, no SO_2 is generated in the fuel cell.

The trend toward portability and miniaturization of computing and communication devices has created a requirement for very small and lightweight power sources that can operate for long periods of time without any refill or replacement. Also, advances in the medical sciences are leading to an increasing number of electrically operated implantable devices like pacemakers, which need power supplies to operate for an extremely long duration (years) without maintenance, as any maintenance would necessitate surgery. Ideally, implanted devices would be able to take advantage of the natural fuel substances found in the body [7–8]. The idea of a biofuel cell that can generate electricity based on various metabolic processes occurring in our own cells is very appealing. A biofuel cell converts chemical energy to electrical energy by the catalytic reaction of microorganisms. Most microbial cells are electrochemically inactive, and electron transfer from microbial cells to the electrode requires mediators such as thionine, methyl viologen, methylene blue, humic acid, and neutral red. In recent years, mediatorless microbial fuel cells have also been developed; these cells use electrochemically active bacteria (*Shewanella putrefaciens*, *Aeromonas hydrophila*, etc.) to transfer electrons to the electrode. A major advantage of the biofuel cell over the hydrogen fuel cell is the replacement of expensive and precious platinum (Pt) as a catalyst by much cheaper hydrogenase enzymes. A brief description of the development and state of the art of hydrogen and biofuel cells is presented in this chapter.

9.2 Fuel Cell Basics

Although fuel cells have been around for more than a century (William Grove in 1839 first discovered the principle of the fuel cell), it was not until the National Aeoronautics and Space Administration (NASA) demonstrated its potential applications in providing power during space flights in the 1960s that fuel cells became widely known and the industry began to recognize the commercial potential of fuel cells. Initially, fuel cells were not economically competitive with existing energy technologies; but with advancements in fuel cell technology, it is now becoming competitive for some niche applications [6].

The main components of a fuel cell are anode, anodic catalyst layer, electrolyte, cathodic catalyst layer, and cathode, as shown in Fig. 9.1. The anode and cathode consist of porous gas diffusion layers, usually made of high-electron-conductivity materials such as thin layers of porous graphite. The most common catalyst is platinum for low-temperature fuel cells. Nickel is preferred for high-temperature fuel

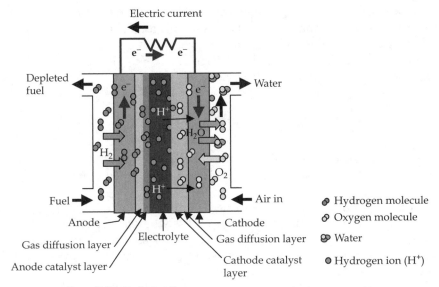

Figure 9.1   Generic H_2-O_2 fuel cell.

cells. Some other materials (Pt-Pt/Ru, Perovskites, etc.) are also used, depending on the fuel cell type [3].

The electrolyte is made up of materials that provide high proton conductivity and zero or very low electron conductivity. The charge carriers (from the anode to the cathode or vice versa) are different, depending on the type of fuel cell. A fuel cell stack is obtained by connecting such fuel cells in series/parallel to yield the desired voltage and current outputs (see Fig. 9.2). The bipolar plates (or interconnects) collect the electrical current and also distribute and separate reactive gases in the fuel cell stack. Sometimes, gaskets for sealing/preventing leakage of gases between anode and cathode are also used.

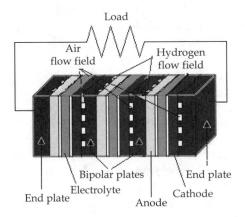

Figure 9.2   A fuel cell stack.

The anode reaction in hydrogen fuel cells is direct oxidation of hydrogen. For fuel cells using hydrocarbon fuels, the anodic half reaction consists of indirect oxidation through a reforming step.

In most fuel cells, the cathode reaction is oxygen (air) reduction. The overall reaction for hydrogen fuel cells is

$$H_2 + \frac{1}{2}O_2 \rightarrow H_2O \qquad \text{with} \qquad \Delta G = -237.2 \text{ kJ/mol}$$

where, ΔG is the change in Gibbs free energy of formation. The reaction product is water released at the cathode or anode, depending on the type of fuel cell.

For an ideal fuel cell, the theoretical voltage E_0 under standard conditions of 25°C and 1 atm pressure is 1.23 V, whereas typical operating voltage for high-performance fuel cells is ~0.7 V. Stack voltage depends on the number of cells in a series in a stack. Cell current depends on the cross-sectional area (the size) of a cell.

Fuel cell systems are not limited by Carnot cycle efficiency. Therefore, a fuel cell system with a combined cycle and/or cogeneration has very high efficiency (55–85%) as compared to the efficiency of about 30–40% of current power generation systems. In a distributed generation system, fuel cells can reduce costly transmission line installation and transmission losses. There are no moving parts in a fuel cell and very few moving parts (compressors, fans, etc.) in a fuel cell system. Therefore, it has higher reliability compared to an internal combustion or gas turbine power plant.

Fuel cell-based power plants have no emissions when pure hydrogen and oxygen are used as fuel. However, if fossil fuels are used for generating hydrogen, fuel cell power plants produce CO_2 emissions. Compared to a steam power plant, a fuel cell plant has very low water usage; water/steam is a reaction product in a fuel cell. This clean water/steam does not require any pretreatment and can be used for reactant humidification and cogeneration. Another advantage of the fuel cell power plant is that it does not produce any solid waste and its operation is very silent as compared to a steam/gas turbine power plant. The noise generated in a fuel cell power plant is only from the fan/compressor used for pumping/pressurizing the fuel and the air supply to the cathode.

A fuel cell power plant has good load-following capability (it can quickly increase or decrease its output in response to load changes). The modular construction of fuel cell plants provides good planning flexibility (new units can be added to meet the growth in electric demand when needed), and its performance is independent of the power plant size (efficiency does not vary with variation in size from W to MW size).

The major technical challenges in fuel cell commercialization at present are (1) high cost, (2) durability, and (3) hydrogen availability and infrastructure. For fuel cells to compete with contemporary power generation technology, they have to become competitive in terms of the cost per kilowatt required to purchase and install a power system. A fuel cell system needs to cost ~$30/kW to be competitive for transportation applications and for stationary systems; the acceptable price range is $400–$750/kW for widespread commercial application [9]. Fuel cell technology needs a few breakthroughs in development to become competitive with other advanced power generation technologies.

9.3   Types of Fuel Cells

Fuel cells are classified primarily on the basis of the electrolyte they use. The electrolyte is the heart of the fuel cell as it decides the important operating parameters such as the electrochemical reactions that take place in the cell, the type of catalysts required, the temperature range of cell operation, and the fuel (reactants) to be used, and therefore the applications for which these cells are most suitable. There are several types of fuel cells currently under development; a few of the most promising types include

- Polymer electrolyte membrane fuel cells (PEMFCs)
- Direct methanol fuel cells (DMFCs)
- Alkaline electrolyte fuel cells (AFCs)
- Phosphoric acid fuel cells (PAFCs)
- Molten carbonate fuel cells (MCFCs)
- Solid oxide fuel cells (SOFCs)
- Biofuel cells

9.3.1   Polymer electrolyte membrane fuel cells (PEMFCs)

The PEMFC uses a solid polymer membrane as an electrolyte. The main components of this fuel cell are an electron-conducting anode consisting of a porous gas diffusion layer as an electrode and an anodic catalyst layer; a proton-conducting electrolyte, a hydrated solid membrane; an electron-conducting cathode consisting of a cathodic catalyst layer and a porous gas diffusion layer as an electrode; and current collectors with the reactant gas flow fields (see Fig. 9.3).

In the PEMFC, platinum or platinum alloys in nanometer-size particles are used as the electrocatalysts with Nafion^{TM} (a DuPont trademark) membranes [3, 10–12]. The polymer electrolyte membranes have some unusual properties: In a hydrated membrane, the negative ions are rigidly held within its structure and are not allowed to pass through.

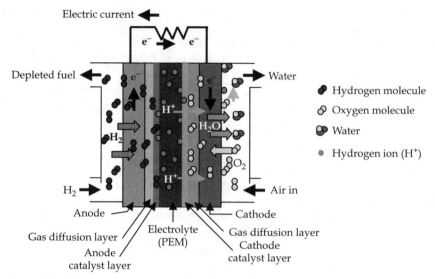

Figure 9.3   Polymer electrolyte membrane fuel cell.

Only the positive ions contained within the membrane are mobile and free to carry a positive charge through the membrane. The proton exchange membrane (PEM) is a good conductor of hydrogen ions (protons), but it does not allow the flow of electrons through the electrolyte membrane. As the electrons cannot pass through the membrane, electrons produced at the anode side of the cell must travel through an external wire to the cathode side of the cell to complete the electrical circuit in the cell.

In the PEMFC, the positive ions moving through the electrolyte are hydrogen ions, or protons. Therefore, the PEMFC is also called a proton exchange membrane fuel cell. The polymer electrolyte membrane is also an effective gas separator; it keeps the hydrogen fuel separated from the oxidant air. This feature is essential for the efficient operation of a fuel cell.

The heart of a PEMFC is the membrane electrode assembly (MEA), consisting of the anode–electrolyte–cathode assembly that is only a few hundred microns thick [11].

Electrochemistry of PEM fuel cells.   All electrochemical reactions consist of two separate reactions: an oxidation half reaction occurring at the anode and a reduction half reaction occurring at the cathode.

Oxidation half reaction:	$2H_2 \rightarrow 4H^+ + 4e^-$
Reduction half reaction:	$O_2 + 4H^+ + 4e^- \rightarrow 2H_2O$
Overall cell reaction:	$2H_2 + O_2 \rightarrow 2H_2O$

The H_2 half reaction. At the anode, hydrogen (H_2) gas molecules diffuse through the porous electrode until they encounter a platinum (Pt) particle. Pt catalyzes the dissociation of the H_2 molecule into two hydrogen atoms (H) bonded to two neighboring Pt atoms; here each H atom releases an electron to form a hydrogen ion (H^+). These H^+ ions move through the hydrated membrane to the cathode while the electrons pass from the anode through the external circuit to the cathode, resulting in a flow of current in the circuit.

The O_2 half reaction. The reaction of one oxygen (O_2) molecule at the cathode is a four-electron reduction process that occurs in a multistep sequence. The catalysts capable of generating high rates of O_2 reduction at relatively low temperatures (~80°C) appear to be the Pt-based expensive catalysts. The performance of the PEMFCs is limited primarily by the slow rate of the O_2 reduction half reaction, which is many times slower than the H_2 oxidation half reaction.

Electrolyte. The polymer electrolyte membrane is a solid organic polymer, usually poly-[perfluorosulfonic] acid. A typical membrane material used in the PEMFC is Nafion [11, 12]. It consists of three regions:

1. The teflon-like fluorocarbon backbone, hundreds of repeating $-CF_2-CF-CF_2-$ units in length
2. The side chains, $-O-CF_2-CF-O-CF_2-CF_2-$, which connect the molecular backbone to the third region
3. The ion clusters consisting of sulfonic acid ions, $SO_3^-\ H^+$

The negative ion SO_3^- is permanently attached to the side chain and cannot move. However, when the membrane becomes hydrated by absorbing water, the hydrogen ion becomes mobile. Ion movement occurs by protons (H^+) bonded to water molecules, hopping from one SO_3^- site to another within the membrane. Because of this mechanism, the solid hydrated electrolyte is an excellent conductor of hydrogen ions.

Electrodes. The anode and the cathode are separated from each other by the electrolyte, the PEM. Each electrode consists of porous carbon to which very small Pt particles are bonded. The porous electrodes allow the reactant gases to diffuse through each electrode to reach the catalyst. Both platinum and carbon are good conductors, so electrons are able to move freely through the electrode [13, 14].

Catalyst. The two half reactions occur very slowly under normal conditions at the low operating temperature (~80°C) of the PEMFC. Therefore, catalysts are needed on both the anode and cathode to increase the rates of each half reaction. Although platinum is a very expensive metal, it is the best material for a catalyst on each electrode.

The half reactions occurring at each electrode can occur only at a high rate at the surface of the Pt catalyst. A unique feature of Pt is that it is sufficiently reactive in bonding H and O intermediates, as required to facilitate the electrode processes, and is also capable of effectively releasing the intermediate to form the final product. The anode process requires Pt sites to bond H atoms when the H_2 molecule reacts; next, these Pt sites release the H atoms, as follows:

$$2H^+ + 2e^- \rightarrow H_2$$

$$H_2 + 2Pt \rightarrow 2(Pt\text{-}H)$$

$$2(Pt\text{-}H) \rightarrow 2Pt + 2H^+ + 2e^-$$

This optimized bonding to H atoms (neither very weak nor very strong) is a unique property of the Pt catalyst. To increase the reaction rate, the catalyst layer is constructed with the highest possible surface area. This is achieved by using very small Pt particles, about 2 nm in diameter, resulting in an enormously large total surface area of Pt that is accessible to gas molecules. The original MEAs for the Gemini space program used 4 mg of platinum per square centimeter of membrane area (4 mg/cm^2). Although the technology varies with the manufacturer, the total platinum loading has decreased from the original 4 mg/cm^2 to about 0.5 mg/cm^2. Laboratory research now uses platinum loadings of 0.15 mg/cm^2. For catalyst layers containing Pt of about 0.15 mg/cm^2, the thickness of the catalyst layer is ~10 μm; the MEA with a total thickness ~200 μm can generate more than half an ampere of current for every square centimeter of the MEA at a voltage of 0.7 V between the cathode and the anode [2, 3, 10–12]. Recently, scientists at Los Alamos National Laboratory, USA have developed a new class of hydrogen fuel cell catalysts that exhibit promising activity and stability. The catalysts, cobalt-polypyrrole-carbon (Co-PPY-XC72) composite, are made of low-cost metals entrapped in a heteroatomic-polymer structure.

The cell hardware. The hardware of the fuel cell consists of backing layers, flow fields, and current collectors. These are designed to maximize the current that can be obtained from an MEA. The backing layers placed next to the electrodes are made of a porous carbon paper or carbon cloth, typically 100–300 μm thick. The porous nature of the backing material ensures effective diffusion of the reactant gases to the catalyst. The backing layers also assist in water management during the operation of the fuel cell; too little or too much water can halt the cell operation. The correct backing material allows the right amount of water vapor to reach the MEA and keep the membrane humidified.

Carbon is used for backing layers because it can conduct the electrons leaving the anode and entering the cathode. A piece of hardware, called a plate, is pressed against the outer surface of each backing layer. The plate serves the dual role of a flow field and current collector. The side of the plate next to the backing layer contains channels machined into the plate. The plates are made of a lightweight, strong, gas-impermeable, electron-conducting material; graphite or metals are commonly used, although composite material plates are now being developed. Electrons produced by the oxidation of hydrogen move through the anode, through the backing layer, and through the plate before they can exit the cell, travel through an external circuit, and reenter the cell at the cathode plate. In a single fuel cell, these two plates are the last of the components making up the cell.

In a fuel cell stack, current collectors are the bipolar plates; they make up over 90% of the volume and 80% of the mass of a fuel cell stack [11, 15, 16].

Water and air management.   Although water is a product of the fuel cell reaction and is carried out of the cell during its operation, it is necessary that both the fuel and air entering the fuel cell be humidified. This additional water keeps the polymer electrolyte membrane hydrated. The humidity of the gases has to be carefully controlled, as too little water dries up the membrane and prevents it from conducting the H^+ ions and the cell current drops. If the air flow past the cathode is too slow, the air cannot carry all the water produced at the cathode out of the fuel cell, and the cathode "floods." Cell performance deteriorates because not enough oxygen is able to penetrate the excess liquid water to reach the cathode catalyst sites. Cooling is required to maintain the temperature of a fuel cell stack at about 80°C, and the product water produced at the cathode at this temperature is both liquid and vapor.

Performance of the PEM fuel cell [3, 11, 16].   Energy conversion in a fuel cell is given by the relation:

Chemical energy of the fuel = electric energy + heat energy

Power is the rate at which energy (E) is made available ($P = dE/dt$, or $\Delta E = P \Delta t$). The power delivered by a cell is the product of the current (I) drawn and the terminal voltage (V) at that current ($P = IV$ watts). In order to compute power delivered by a fuel cell, we have to know the cell voltage and load current. The ideal (maximum) cell voltage (E) for the hydrogen/air fuel cell reaction ($H_2 + 1/2O_2 \rightarrow H_2O$) at a specific temperature and pressure is calculated from the maximum electrical energy

$$W_{el} = -\Delta G = nFE \quad \text{or} \quad E = -\frac{\Delta G}{nF}$$

where ΔG is the change in Gibbs free energy for the reaction, n is the number of moles of electrons involved in the reaction per mole of H_2, and F (Faraday's constant) = 96,487 C (coulombs = joules/volt). At a constant pressure of 1 atm, the change in Gibbs free energy in the fuel cell process (per mole of H_2) is calculated from the reaction temperature (T) and from changes in the reaction enthalpy (H) and entropy (S).

$$\Delta G = \Delta H - T\,\Delta S$$
$$= -285{,}800 \text{ J} - (298 \text{ K})(-163.2 \text{ J/K})$$
$$= -237{,}200 \text{ J}$$

For the hydrogen–air fuel cell at 1 atm pressure and 25°C (298 K), the cell voltage is

$$E = -\frac{\Delta G}{nF}$$
$$= -\left(-\frac{237{,}200 \text{ J}}{2 \times 96{,}487 \text{ J/V}}\right) = 1.23 \text{ V}$$

As temperature rises from room temperature to the PEM fuel cell operating temperature (80°C or 353 K), the change in values of H and S is very small, but T changes by 55°C. Thus the absolute value of ΔG decreases. Assuming negligible change in the values of H and S,

$$\Delta G = -285{,}800 \text{ J/mol} - (353 \text{ K})(163.2 \text{ J/mol} \cdot \text{K})$$
$$= -228{,}200 \text{ J/mol}$$

Therefore,

$$E = -\left(-\frac{228{,}200 \text{ J}}{2 \times 96{,}487 \text{ J/V}}\right) = 1.18 \text{ V}$$

Thus, for standard pressure of 1 atm, the maximum cell voltage decreases from 1.23 V at 25°C to 1.18 V at 80°C. An additional correction is needed for using air instead of pure oxygen, and also for using humidified air and hydrogen instead of dry gases. This further reduces the maximum voltage from the hydrogen–air fuel cell to 1.16 V at 80°C and 1 atm pressure. With an increase in load current, the actual cell potential is decreased from its no-load potential because of irreversible losses, which are often called polarization or overvoltage (h). These originate primarily from three sources:

- Activation polarization (h_{act})
- Ohmic polarization (h_{ohm})
- Concentration polarization (h_{conc})

The polarization losses result in a further decrease in actual cell voltage (V) from its ideal potential E ($V = E -$ potential drop due to losses). The activation polarization loss is dominant at low current density. This is because electronic barriers have to be overcome prior to current and ion flows. Activation polarization is present when the rate of an electrochemical reaction at an electrode surface is controlled by sluggish electrode kinetics. Therefore, activation polarization is directly related to the rates of electrochemical reactions. In an electrochemical reaction with $h_{act} > 50 - 100$ mV, activation polarization is described by a semi-empirical equation known as the Tafel equation:

$$h_{act} = \left(\frac{RT}{\alpha nF}\right) \ln\left(\frac{i}{i_0}\right)$$

where α is the electron transfer coefficient of the reaction at the electrode (anode or cathode), and i_0 is the exchange current density. The Tafel slope for the PEMFC electrochemical reaction is about 100 mV/decade at room temperature. Thus there is an incentive to develop electrocatalysts that yield a lower Tafel slope [2, 3, 11, 12].

Ohmic losses occur because of the resistance to the flow of ions in the electrolyte and resistance to the flow of electrons through the electrode materials. Decreasing the electrode separation and enhancing the ionic conductivity of the electrolyte can reduce the ohmic losses. Both the electrolyte and fuel cell electrodes obey Ohm's law; the ohmic losses can be expressed by the equation: $h_{ohm} = iR$, where i is the current flowing through the cell and R is the total cell resistance, which includes ionic, electronic, and contact resistance. See Figure 9.4.

Due to the consumption of reactants at the electrode by an electrochemical reaction, the surrounding material is unable to maintain the

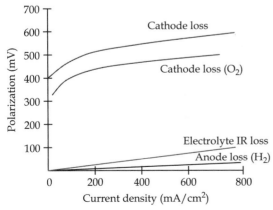

Figure 9.4 Activation losses in a PEM fuel cell [1].

initial concentration of the bulk fluid and a concentration gradient is formed, resulting in a loss of electrode potential. Although several processes contribute to concentration polarization, at practical current densities, slow transport of reactants and products to and from the electrochemical reaction site is a major contributor to concentration polarization. The effect of polarization is to shift the potential of the electrode:

For the anode,

$$V_{\text{anode}} = E_{\text{anode}} + |h_{\text{anode}}|$$

and for the cathode,

$$V_{\text{cathode}} = E_{\text{cathode}} - |h_{\text{cathode}}|$$

The net result of current flow in a fuel cell is to increase the anode potential and to decrease the cathode potential. This reduces the cell voltage. The cell voltage includes the contribution of the anode and cathode potentials and ohmic polarization. See Figure 9.5.

$$V_{\text{cell}} = V_{\text{cathode}} - V_{\text{anode}} - iR; \text{ or}$$

$$V_{\text{cell}} = E_{\text{cathode}} - |h_{\text{cathode}}| - (E_{\text{anode}} + |h_{\text{anode}}|) - iR; \text{ or}$$

$$V_{\text{cell}} = E_{\text{cell}} - |h_{\text{cathode}}| - |h_{\text{anode}}| - iR$$

where $E_{\text{cell}} = E_{\text{cathode}} - E_{\text{anode}}$

The goal of fuel cell developers is to minimize the polarization losses so that the V_{cell} approaches the E_{cell} by modifications to the fuel cell

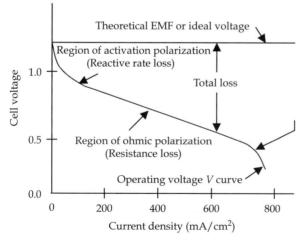

Figure 9.5 PEM fuel cell voltage versus current density curve [3].

design by improvement in the electrode structures, better electrocatalysts, more conductive electrolytes, thinner cell components, and so forth. It is possible to improve the cell performance by modifying the operating conditions such as higher gas pressure, higher temperature, and a change in gas composition to lower the gas impurity concentration [3].

9.3.2 Direct methanol fuel cells (DMFCs)

Direct methanol fuel cells are similar to the PEMFC as they also use a polymer membrane as the electrolyte. However, it produces power by direct conversion of liquid methanol to hydrogen ions on the anode side of the fuel cell. In the DMFC, the anode catalyst draws hydrogen directly from the liquid methanol, thus eliminating the need for a fuel reformer. All the DMFC components (anode, cathode, membrane, and catalysts) are the same as those of a PEMFC. A DMFC system is shown in Fig. 9.6. Methanol diluted to a specified concentration is fed to the fuel cell stack. During operation, the concentration of the methanol solution exiting the stack is reduced. Therefore, pure methanol is added in the feed cycle to restore the original concentration of the solution. A gas–liquid separator is used to remove carbon dioxide from the solution loop, and a compressor feeds air to the DMFC stack. Water and heat are recovered by passing the outlet air through a condenser. A portion of the recovered water is returned to the fuel circulation loop. The stack temperature is maintained by removing the excess heat from the fuel circulation loop using a heat exchanger. The DMFC can attain high efficiencies of 40% with a Nafion-117 membrane at 60°C, with current

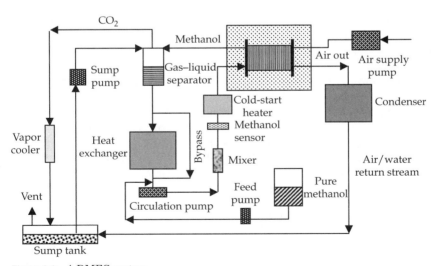

Figure 9.6 A DMFC system.

density in the range of 100–120 mA/cm^2. Studies have shown that DMFC efficiency decreases with increasing methanol concentration. Therefore, operating a fuel cell to maintain the maximum efficiency needs close control of methanol concentration and temperature. An online concentration sensor is used in the feedback loop for this purpose. Some of the advantages of this system, relative to the hydrogen systems, are that the liquid feed (methanol) helps in attaining the uniform stack temperature and maintenance of membrane humidity; it is also easy to refill since the fuel (methanol) is in liquid form.

As compared to the PEMFC, the DMFC has a very sluggish electrochemical reaction (significant activation over voltage) at the anode. It therefore requires a high surface area of 50:50% Pt-Ru (a more expensive bimetal) alloy as the anode catalyst to overcome the sluggish reaction and an increase in catalyst loading of more than 10 times that for the PEMFC. Even then, the output voltage on the load is only 0.2–0.4 V with an efficiency of about 40% at operating temperatures between 60°C and 90°C. This is relatively low, and therefore, the DMFC is attractive only for tiny to small-sized applications (cellular phones, laptops, etc.) [17]. Another potential application for the DMFC is in transport vehicles; as it operates on liquid fuels, it would greatly simplify the onboard system as well as the infrastructure needed to supply fuel to passenger cars and commercial fleets and can create a large potential market for commercialization of fuel cell technology in vehicle applications.

9.3.3 Alkaline-electrolyte fuel cells (AFCs)

Alkaline-electrolyte fuel cells (see Fig. 9.7) are one of the most developed fuel cell technologies. They have been in use since the mid-1960s for Apollo and space shuttle programs [3, 6, 18, 19]. The AFCs onboard these spacecraft provide electrical power as well as drinking water. AFCs are among the most efficient electricity-generating fuel cells with an efficiency of nearly 70%. The electrolyte used in the AFC is an alkaline solution in which an OH$^-$ ion can move freely across the electrolyte.

Electrochemistry of AFCs.   The electrolyte used in the AFC is an aqueous (water-based) solution of potassium hydroxide (KOH) retained in a porous stabilized matrix. The concentration of KOH can be varied with the fuel cell operating temperature, which ranges from 65 to 220°C.

The charge carrier for an AFC is the hydroxyl ion (OH$^-$) that migrates from the cathode to the anode, where they react with hydrogen to produce water and electrons. Water formed at the anode migrates back to the cathode to regenerate hydroxyl ions.

Anode reaction: $2H_2 + 4OH^- \rightarrow 4H_2O + 4e^-$

Cathode reaction: $O_2 + 2H_2O + 4e^- \rightarrow 4OH^-$

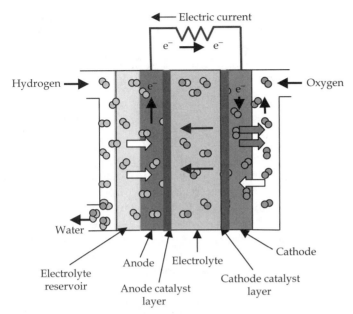

Figure 9.7   An alkaline-electrolyte fuel cell.

Hydroxyl ions are the conducting species in the electrolyte.

Overall cell reaction: $2H_2 + O_2 \rightarrow 2H_2O$ + heat + electricity

In many cell designs, the electrolyte is circulated (mobile electrolyte) so that heat can be removed and water eliminated by evaporation. Since KOH has the highest conductance among the alkaline hydroxides, it is the preferred electrolyte.

Electrolyte. Concentrated KOH (85 wt.%) is used in cells designed for operation at a high temperature (~260°C). For lower temperature (<120°C) operation, less concentrated KOH (35–50 wt.%) is used. The electrolyte is retained in a matrix (usually asbestos), and a wide range of electrocatalysts can be used (e.g., Ni, Ag, metal oxides, and noble metals). A major advantage of the AFC is the lower activation polarization at the cathode, resulting in a higher operating voltage (0.875 V). Another advantage of the AFC is the use of inexpensive electrolyte materials. The electrolyte is replenished through a reservoir on the anode side. The typical performance of this AFC cell is 0.85 V at a current density of 150 mA/cm^2. The AFCs used in the space shuttle orbiter have a rectangular cross-section and weigh 91 kg. They operate at an average power of 7 kW with a peak power rating of 12 kW at 27.5 V. A disadvantage of the AFC is that it is very sensitive to CO_2 present in the fuel or air. The alkaline electrolyte reacts with CO_2 and severely degrades the fuel cell performance, limiting their application to closed environments, such as space and undersea vehicles, as these cells work well only with pure hydrogen and oxygen as fuel.

Electrodes. A significant cost advantage of alkaline fuel cells is that both anode and cathode reactions can be effectively catalyzed with non-precious, relatively inexpensive metals. The most important characteristics of the catalyst structure are high electronic conductivity and stability (mechanical, chemical, and electrochemical). Both metallic (typically hydrophobic) and carbon-based (typically hydrophilic) electrode structures with multilayers and optimized porosity characteristics for the flow of liquid electrolytes and gases (H_2 and O_2) have been developed. The kinetics of oxygen reduction in alkaline electrolytes is much faster than in acid media; hence AFCs can use low-level Pt catalysts (about 20% Pt, compared with PEMFCs) on a large surface carbon support [20].

Performance. The AFC development has gone through many changes since 1960. To meet the requirements for space applications, the early AFCs were operated at relatively high temperatures and pressures. Now the focus of the technology is to develop low-cost components for AFCs operating at near-ambient temperature and pressure, with air as the oxidant for terrestrial applications. This has resulted in lower performance. The reversible cell potential for an H_2 and O_2 fuel cell decreases by 0.49 mV/°C under standard conditions. An increase in operating temperature reduces activation polarization, mass transfer polarization, and ohmic losses, thereby improving cell performance. Alkaline cells operated at low temperatures (~70°C) show reasonable performance.

Pure hydrogen and oxygen are required in order to operate an AFC. Reformed H_2 or air containing even trace amounts of CO_2 dramatically affects its performance and lifetime. There is a drastic loss in performance when using hydrogen-rich fuels containing even a small amount of CO_2 from reformed hydrocarbon fuels and also from the presence of CO_2 in the air (~350 ppm CO_2 in ambient air). The CO_2 reacts with OH^- (CO_2 + $2OH^- \rightarrow CO_3^{2-} + H_2O$), thereby decreasing their concentration and thus reducing the reaction kinetics. Other ill effects of the presence of CO_2 are:

- Increase in electrolyte viscosity, resulting in lower diffusion rate and lower limiting currents.
- Deposition of carbonate salts in the pores of the porous electrode.
- Reduction in oxygen solubility.
- Reduction in electrolyte conductivity.

A higher concentration of KOH decreases the life of O_2 electrodes when operating with air containing CO_2. However, operation at higher temperatures is beneficial because it increases the solubility of CO_2 in the electrolyte. The operational life of air electrodes polytetrafluoroethylene [PTFE] bonded carbon electrodes on porous nickel substrates) at a current density of 65 mA/cm^2 in 9-N KOH at 65°C ranges from 4000 to 5500 h with CO_2-free air, and their life decreases to 1600–3400 h when air (350-ppm CO_2) is used. For large-scale utility applications, operating times >40,000 h are required, which is a very significant hurdle to commercialization of AFC devices for stationary electric power generation.

Another problem with the AFC is that the electrodes and catalysts degrade more on no-load or light-load operation than on a loaded condition, because the high open-circuit voltage causes faster carbon oxidation processes and catalyst changes. The AFC with immobilized KOH electrolyte suffers much more from this as the electrolyte has to stay in the cells causing residual carbonate accumulation, separator deterioration, and gas cross leakage during storage or unloaded periods if careful maintenance is not carried out. In circulating an electrolyte-type AFC, the electrolyte is emptied from the cell during nonoperating periods. Shutting off the H_2 electrodes from air establishes an inert atmosphere. This shutdown also eliminates all parasitic currents and increases life expectancy. The exchangeability of the KOH in a circulating electrolyte-type AFC offers the possibility to operate on air without complete removal of the CO_2 [20, 21].

9.3.4 Phosphoric acid fuel cells (PAFCs)

Phosphoric acid fuel cells (see Fig. 9.8) operate at intermediate temperatures (~200°C) and are very well developed and commercially available today. Hundreds of PAFC systems are working around the world in hospitals, hotels, offices, schools, utility power plants, landfills and wastewater treatment plants, and so forth. Most of the PAFC plants are in the 50- to 200-kW capacity ranges, but large plants of 1- and 5-MW capacity have also been built; a demonstration unit has achieved 11 MW of grid-quality ac power [3]. PAFCs generate electricity at more than 40% efficiency and if the steam produced is used for cogeneration, efficiencies of nearly 85% can be achieved. PAFCs use liquid phosphoric acid as the electrolyte. One of the main advantages to this type of fuel cell, besides high efficiency, is that it does not require pure hydrogen as fuel and can tolerate up to 1.5% CO concentration in fuel, which broadens the choice of fuels that can be used. However, any sulfur compounds present in the fuel have to be removed to a concentration of <0.1 ppmV. Temperatures of about 200°C and acid concentrations of 100% H_3PO_4 are commonly used, while operating pressure in excess of 8 atm has been used in an 11-MW electric utility demonstration plant [3, 22, 23].

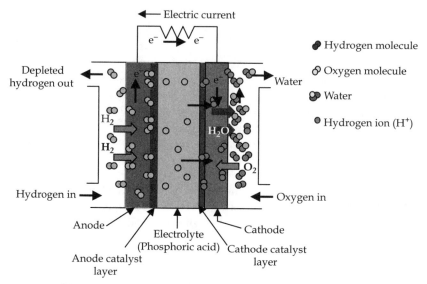

Figure 9.8   Phosphoric acid fuel cell.

Electrochemistry of PAFCs.   The electrochemical reactions occurring in a PAFC are

At the anode:

$$H_2 \rightarrow 2H^+ + 2e^-$$

At the cathode:

$$\frac{1}{2}O_2 + 2H^+ + 2e^- \rightarrow H_2O$$

The overall cell reaction:

$$\frac{1}{2}O_2 + H_2 \rightarrow H_2O$$

The fuel cell operates on H_2; CO is a poison when present in a concentration greater than 0.5%. If a hydrocarbon such as natural gas is used as a fuel, reforming of the fuel by the reaction

$$CH_4 + H_2O \rightarrow 3H_2 + CO$$

and shifting of the reformat by the reaction

$$CO + H_2O \rightarrow H_2 + CO_2$$

is required to generate the required fuel for the cell.

Electrolyte.   The PAFC uses 100% concentrated phosphoric acid (H_3PO_4) as an electrolyte. The electrolyte assembly is a 0.1- to 0.2-mm-thick matrix made of silicon carbide particles held together with a small amount of PTFE. The pores of the matrix retain the electrolyte (phosphoric acid) by capillary action. At lower temperatures, H_3PO_4 is a poor ionic conductor and CO poisoning of the Pt electrocatalyst in the anode can become severe. There will be some loss of H_3PO_4 over long periods, depending upon the operating conditions. Hence, as a general rule, sufficient acid reserve is kept in the matrix at the beginning.

Electrode.   The PAFC (similar to a PEMFC) uses gas diffusion electrodes. Platinum or platinum alloys are used as the catalyst at both electrodes. In the mid-1960s, the conventional porous electrodes were PTFE-bonded Pt black, and the loadings of Pt were about 9 mg/cm^2. In recent years, Pt supported on carbon black has replaced Pt black in porous PTFE-bonded electrode structures. Pt loading has also dramatically reduced to about 0.25 mg Pt/cm^2 in the anode and about 0.50 mg Pt/cm^2 in the cathode. The porous electrodes used in a PAFC consist of a mixture of the electrocatalyst supported on carbon black and a polymeric binder to bind the carbon black particles together to form an integral structure. A porous carbon paper substrate provides structural support for the electrocatalyst layer and also acts as the current collector. The composite structure consisting of a carbon black/binder layer onto the carbon paper substrate forms a three-phase interface, with the electrolyte on one side and the reactant gases on the other side of the carbon paper. The stack consists of a repeating arrangement of a bipolar plate, the anode, electrolyte matrix, and cathode.

Hardware.   A bipolar plate separates the individual cells and electrically connects them in a series in a fuel cell stack. A bipolar plate has a multifunction design; it has to separate the reactant gases in the adjacent cells in the stack, so it must be impermeable to reactant gases; it must transmit electrons to the next cell (series connection), so it has to be electrically conducting; and it must be heat conducting for proper heat transfer and thermal management of the fuel cell stack. In some designs, gas channels are also provided on the bipolar plates to feed reactant gases to the porous electrodes and to remove the reaction products. Bipolar plates should have very low porosity so as to minimize phosphoric acid absorption. These plates must be stable and corrosion-resistant in the PAFC environment. Bipolar plates are usually made of graphite–resin mixtures that are carbonized and heat treated to 2700°C to increase corrosion resistance. For 100-kW and larger power generation systems, water cooling has to be used and cooling channels are provided in the bipolar plates to cool the stack.

Temperature and humidity management. Temperature and humidity management are essential for proper operation of a PAFC. The PAFC system has to be heated up to 130°C before the cell can start working. At lower temperatures, concentrated phosphoric acid does not get dissociated, resulting in a low availability of protons. Also, due to lower vapor pressure of the concentrated acid, the water generated will not come out with the reactant stream and the moisture retention dilutes the acid. This causes an increase in acid volume, which results in acid oozing out through the electrode. With the start of normal cell operation, its temperature increases and acid concentration gets back to its normal value that causes acid volume to shrink, resulting in drying of the electrolyte matrix pores if the acid is not replenished. Controlled stack heating at start-up is achieved by using an insertable heater system. During operation, the temperature of the stack is maintained by controlling the air flow in the oxidant channel. At high loading conditions, insertable coolers may be used to remove excess heat from the stack. Large-power PAFC systems use a water-cooling system.

Moisture generated at the cathode dilutes the acid on the cathode side of the electrolyte matrix, causing higher vapor pressure. This results in more moisture out with the oxidant stream. With the movement of protons from anode to cathode, moisture migration takes place at the cathode side also. This water evaporation results in an acid concentration gradient from anode to cathode, causing low availability of protons and a lower potential of the cell. Therefore, water management is needed to maintain humidity of the anode stream gas at a sufficient level so that the vapor pressure matches the acid concentration level at the operating temperature.

Performance. For good performance, the normal operating temperature range of a PAFC is $180°C < T < 250°C$; below 200°C, the decrease in cell potential is significant. Although an increased temperature increases performance, higher temperatures also result in increased catalyst sintering, component corrosion, electrolyte degradation, and evaporation. PAFCs operate in the current density range of 100–400 mA/cm^2 at 600–800 mV/cell. Voltage and power limitations result from increased corrosion of platinum and carbon components at cell potentials above approximately 800 mV. Since the freezing point of phosphoric acid is 42°C, the PAFC must be kept above this temperature once commissioned to avoid the thermal stresses due to freezing and thawing. Various factors affect the PAFC life. Acid concentration management by proper humidity control is very important to prevent acid loss and performance degradation. A PAFC has a life of 10,000–50,000 h, commercially available (UTC Fuel Cells) PAFC systems operating at 207°C have shown a life of 40,000 h with reasonable performance (degradation rate $\Delta V_{lifetime}$ (mV) $= -2$ mV/1000 h) [3, 23].

9.3.5 Molten carbonate fuel cells (MCFCs)

The MCFC has evolved from work in the 1960s, aimed at producing a fuel cell that would operate directly on coal [23, 24]. Although direct operation on coal is no longer a goal, a remarkable feature of the MCFC is that it can directly operate on coal-derived fuel gases or natural gas and is therefore also called a direct fuel cell (DFC). MCFCs operate at high temperatures (600–650°C) compared to phosphoric acid (180–220°C) or PEM fuel cells (60–85°C). Operation at high temperatures eliminates the need for external fuel processors that the lower temperature fuel cells require to extract hydrogen from naturally available fuel. When natural gas is used as fuel, methane (the main ingredient of natural gas) and water (steam) are converted into a hydrogen-rich gas inside the MCFC stack ("internal reforming") (see Fig. 9.9). High operating temperatures also result in high-temperature exhaust gas, which can be utilized for heat recovery for secondary power generation or cogeneration. MCFCs can therefore achieve a higher fuel-to-electricity and an overall energy use efficiency (>75%) than the low-temperature fuel cells. The MCFC

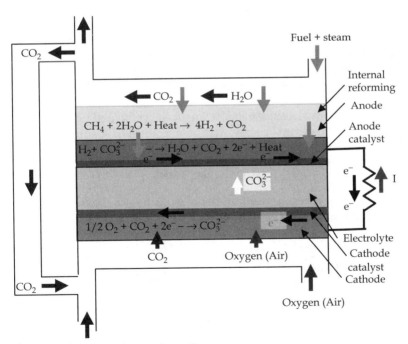

Figure 9.9   Molten carbonate fuel cell.

is a well-developed fuel cell and is a commercially viable technology for a stationary power plant, compared to other fuel cell types. A number of MCFC prototype units in the power range of 200 kW to 1 MW and higher are operating around the world. The cost and useful life issues are the major challenges to overcome before the MCFC can compete with the existing (thermal or other) electric power generation systems for widespread use.

Electrochemistry of MCFC.   The electrochemical reactions occurring in the cell are:

Anode half reaction. At the anode, hydrogen reacts with carbonate ions to produce water, carbon dioxide, and electrons. The electrons travel through an external circuit—creating electricity—and return to the cathode.

$$H_2 + CO_3^{2-} \rightarrow H_2O + CO_2 + 2e^-$$

Cathode half reaction. At the cathode, oxygen from the air and carbon dioxide recycled from the anode react with the electrons to form carbonate ions that replenish the electrolyte and transfer the current through the fuel cell, completing the circuit.

$$\frac{1}{2}O_2 + CO_2 + 2e^- \rightarrow CO_3^-$$

The overall cell reaction is

$$H_2 + \frac{1}{2}O_2 + CO_2 \text{ (cathode)} \rightarrow H_2O + CO_2 \text{ (anode)}$$

If a fuel such as natural gas is used, it has to be reformed either externally or within the cell (internally) in the presence of a suitable catalyst to form H_2 and CO by the reaction:

$$CH_4 + H_2O \rightarrow 3H_2 + CO$$

Although, CO is not directly used by the electrochemical oxidation, but produces additional H_2 by the water gas shift reaction:

$$CO + H_2O \rightarrow H_2 + CO_2$$

Typically, the CO_2 generated at the anode is recycled to the cathode, where it is consumed. This requires additional equipment to either transfer CO_2 from the anode exit gas to the cathode inlet gas or produce CO_2 by combustion of anode exhaust gas and mix with the cathode inlet gas.

Electrolyte. The MCFC uses a molten carbonate salt mixture as its electrolyte. At operating temperatures of about 650°C, the salt mixture is in a molten (liquid) state and is a good ionic conductor. The composition of salts in the electrolyte may vary but usually consist of lithium/potassium carbonate (Li_2CO_3/K_2CO_3, 62–38 mol%) for operation at atmospheric pressure. For operation under pressurized conditions, lithium/sodium carbonate ($LiCO_3/NaCO_3$, 52–48 or 60–40 mol%) is used as it provides improved cathode stability and performance. This allows for the use of thicker Li/Na electrolyte for the same performance, resulting in a longer lifetime before a shorting caused by internal precipitation. The composition of the electrolyte has an effect on electrochemical activity, corrosion, and electrolyte loss rate. Li/Na offers better corrosion resistance but has greater temperature sensitivity. Additives are being developed to minimize the temperature sensitivity of the Li/Na electrolyte. The electrolyte has a low vapor pressure at the operating temperature and may evaporate very slowly; however, this does not have any serious effect on the cell life. The electrolyte is suspended in a porous, insulating, and chemically inert ceramic ($LiAlO_2$) matrix. The ceramic matrix has a significant effect on the ohmic resistance of the electrolyte. It accounts for almost 70% of the ohmic polarization. The electrolyte management in an MCFC ensures that the electrolyte matrix remains completely filled with the molten carbonate, while the porous electrodes are partially filled, depending on their pore size distributions.

Electrode. The anode is made of a porous chromium-doped sintered Ni-Cr/Ni-Al alloy. Because of the high temperatures resulting in a fast anode action, a large surface area is not required on the anode as compared to the cathode. Partial flooding of the anode with molten carbonate is desirable as it acts as a reservoir that replenishes carbonate in the stack during prolonged use. The cathode is made up of porous lithiated nickel oxide. Because of the high operating temperatures, no noble catalysts are needed in the fuel cell. Nickel is used on the anode and nickel oxide on the cathode as catalysts. Bipolar plates or interconnects are made from thin stainless steel sheets with corrugated gas diffusion channels. The anode side of the plate is coated with pure nickel to protect against corrosion.

Performance. At the high operating temperatures of an MCFC, CO is not a poison but acts as a fuel. In the MCFC, CO_2 has to be added to oxygen (air) stream at the cathode for generation of carbonate ions. The anode reaction converts these ions back to CO_2, resulting in a net transfer of two ions with every molecule of CO_2. The need for CO_2 in the oxidant stream requires that CO_2 from the spent anode gas be separated and mixed with the incoming air stream. Before this can be done, any residual hydrogen in the spent fuel stream must be burned. Systems developed in the future may incorporate membrane separators to remove the hydrogen for recirculation back to the fuel stream to increase efficiency.

Internal reforming of natural gas and partially cracked hydrocarbons is possible in the inlet chamber of the MCFC, eliminating the separate fuel processing of natural gas or other hydrogen-rich fuels. The requirement for CO_2 makes the digester gas (sewage, animal waste, food processing waste, etc.) an ideal fuel for the MCFC; other fuels such as natural gas, landfill gas, propane, coal gas, and liquid fuels (diesel, methanol, ethanol, LPG, etc.) can also be used in the MCFC system. The elimination of the external fuel reformer also contributes to lower costs, and high-temperature waste heat can be utilized to make additional electricity and cogeneration. MCFCs can reach overall thermal efficiencies as high as 85%.

With the increase in operating temperature, the theoretical operating voltage for a fuel cell decreases, but increases the rate of the electrochemical reaction and therefore the current that can be obtained at a given voltage. This results in the MCFC having a higher operating voltage for the same current density and higher fuel efficiency than a PAFC of the same electrode area. As size and cost scale roughly with the electrode area, the MCFC is smaller and less expensive than a PAFC of comparable output. Another advantage of the MCFC is that the electrodes can be made with cheaper nickel catalysts rather than the more expensive platinum used in other low-temperature fuel cells. Endurance of the cell stack is a critical issue in commercialization, and MCFC manufacturers report an average potential degradation of -2 mV/1000 h over a cell stack lifetime of 40,000 h. The high temperature limits the use of materials in the MCFC, and safety issues prevent their application for home use. MCFC units require a few minutes of fuel burning at the start up to heat up the cell to its operating temperature and therefore are not very suitable for use in automobiles. However, they are very good for stationary power applications and units with up to 2 MW have been constructed, and designs for units with up to 100 MW exist [3, 23–25].

9.3.6   Solid oxide fuel cells (SOFCs)

The SOFC has the most desirable properties for generating electricity from hydrocarbon fuels. The SOFC uses a solid electrolyte and is very efficient. It can internally reform hydrocarbon fuels and is tolerant to impurities. The SOFC operates at a very high temperature (700–1000°C) and so does not require any cooling system for maintaining a fuel cell operating temperature. For small systems, insulation has to be provided to maintain the cell temperature. In large SOFC systems, the operating temperature is maintained internally by the reforming action of the fuel and by the cool outside air (oxidant) that is drawn into the fuel cell. At high operating temperatures, chemical reaction rates in the SOFC are high and air compression is not required. This results in a simpler system, quiet operation, and high efficiencies. Westinghouse has worked at developing a tubular style of the SOFC that operates at 1000°C (see Fig. 9.10) for many years [1–3, 26, 27]. These long tubes have high electrical resistance but are simple to seal. Many other manufacturers are now working on a planar SOFC composed of thin ceramic sheets which operate at 800°C or even less. Thin sheets offer low electrical resistance, and cheaper materials such as stainless steel can be used at these lower temperatures [3, 6, 26]. One big advantage of the SOFC over the MCFC is that the electrolyte is a solid. Therefore, no pumps are required to circulate a hot electrolyte, and very compact, small planar SOFC systems of a few kW range could be constructed using very thin sheets.

A major advantage of the SOFC is that both hydrogen and carbon monoxide are used in the cell. Therefore, in the SOFC, many common hydrocarbon fuels such as natural gas, diesel, gasoline, alcohol, and coal gas can be safely used. The SOFC can reform these fuels into hydrogen

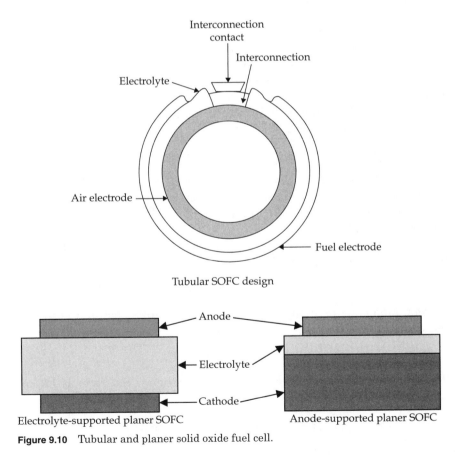

Tubular SOFC design

Electrolyte-supported planer SOFC   Anode-supported planer SOFC

Figure 9.10   Tubular and planer solid oxide fuel cell.

and carbon monoxide inside the cell, and the high-temperature waste thermal energy can be recycled back for fuel reforming. During operation, the SOFC is at the same time a generator and a user of heat. Heat is generated through exothermic chemical reactions and ohmic losses, while it is absorbed by the reforming reaction. It is possible to design the SOFC to be thermally balanced, thereby eliminating the requirement for external insulation and heating. Small SOFC systems are not thermally self-sustaining and may require an external heat source to start and maintain operation. In large systems, the heat generated is not fully absorbed by fuel reforming, and the excess heat can be used in gas turbines for generating electricity or for cogeneration. Another advantage of the SOFC is that expensive catalysts are not required. However, a few minutes of fuel burning is required to reach the operating temperature of the SOFC at the start. This time delay is a disadvantage for an automotive application, but for stationary electric power plants, this is not a problem as they run continuously for long periods of time.

Electrochemistry of SOFCs.   Hydrogen or carbon monoxide in the fuel stream reacts with oxide ions (O^{2-}) from the electrolyte to produce water or CO_2 and to deposit electrons into the anode. The electrons pass outside the fuel cell, through the load, and back to the cathode, where oxygen from the air receives the electrons and is converted into oxide ions, which are injected into the electrolyte. In the SOFC, oxygen ions are formed at the cathode. The reaction at the cathode is

$$O_2 + 4e^- \rightarrow 2O^{2-}$$

At the operating temperature, the electrolyte offers high ionic conductivity and low electrical conductivity; therefore, oxygen ions migrate through the electrolyte to the anode. The overall reaction occurring at the anode is as follows:

The hydrogen in the fuel reacts with the oxygen ions to produce water and releases two electrons.

$$H_2 + O^{2-} \rightarrow H_2O + 2e^-$$

Carbon monoxide present in the fuel causes a shift reaction to produce additional fuel (H_2).

$$CO + H_2O \leftrightarrow H_2 + CO_2$$

The following internal reforming reaction for the hydrocarbon fuel takes place on the anode side:

$$C_xH_y + xH_2O \rightarrow xCO + (x + \frac{y}{2})H_2$$

For methane-rich fuels, this reforming reaction is

$$CH_4 + H_2O \rightarrow CO + 3H_2$$

This reaction is generally not in chemical equilibrium, and the CO shift reaction takes place to provide more hydrogen. The overall cell reaction is

$$H_2 + \frac{1}{2}O_2 \rightarrow H_2O$$

Electrolyte. The use of a solid electrolyte in the SOFC eliminates the electrolyte management problems associated with the liquid electrolyte fuel cells and also reduces corrosion considerations to a great extent. In the SOFC, it is the migration of oxygen ions (O^{2-}) through the electrolyte that establishes the voltage difference between the anode and cathode. Therefore, the electrolyte must be a good conductor of O^{2-} ions and a bad conductor of electrons; it must also be stable at the high operating temperature. Some ceramics possess these properties and therefore are good candidates for this application. With the help of modern ceramic technology and solid-state science, many ceramics can be tailored for electrical properties unattainable in metallic or polymer materials. These tailored ceramic materials are termed electroceramics, and one group is known as *fast ion conductors* or *superionic conductors*. These superionic conductors when used as a solid electrolyte allow easy passage of ions from the cathode to the anode in an SOFC. The material generally used as an electrolyte in the SOFC is dense yttria-stabilized zirconia. It is an excellent conductor of negatively charged oxygen ions at high temperatures (1000°C), but its conductivity reduces drastically with the drop in temperature. Other materials such as scandia-stabilized zirconia (ScSZ), which shows good ionic conductivity at a lower temperature (800°C), are also being investigated, but the electrolyte developed with ScSZ-based materials is very expensive and they degrade very fast.

Electrode. The anode is made of metallic Ni and Y_2O_3-stabilized ZrO_2 (YSZ). Yttria-stabilized zirconia is added in Ni to inhibit sintering of the metal particles and to provide a thermal expansion coefficient close to those of the other cell materials [26]. Nickel structure is normally obtained from NiO powders; therefore, before starting the operation for the first time, the cell is run with hydrogen in an open-circuit condition to reduce the NiO to nickel. The anode structure is fabricated with a porosity of 20–40% to facilitate mass transport of the reactant and product gases. The Sr-doped lanthanum manganite ($La_{1-x}Sr_xMnO_3$, $x = 0.10$–0.15; known as LSM) is most commonly used for the cathode material. LSM is a p-type semiconductor. Similar to the anode, the cathode is also a porous structure that permits rapid mass transport of the reactant and product gases.

Hardware. In the SOFC, both CO and hydrogen are used as direct fuel. Therefore, it is important that the fuel and air streams are kept separate, and a thermal balance should be maintained to ensure that operating temperatures remain within an acceptable range. Several designs of the SOFC (tubular and planer) have been developed to accommodate these requirements. The SOFC is a solid-state device and shares certain properties and fabrication techniques with semiconductor devices.

Individual cells in the stack are connected by interconnects, which carry an electrical current between cells and can also act as a separator between the fuel and oxidant supplies. In high-temperature SOFCs, the interconnects that are used are ceramic such as lanthanum chromite, or if the temperature is limited to less than 1000°C, a refractory alloy based on Y/Cr may be used. The interconnects constitute a major proportion of the stack cost. Stack and other plant construction materials that are used also need to be refractory to withstand the high-temperature gas streams. Volatility of chromium-containing ceramics and alloys can result in contamination of the stack components, and the presence of a toxic material such as Cr^{6+} requires special disposal procedures.

The high operating temperature (1000°C) of the SOFC requires a significant start-up time. The cell performance is very sensitive to operating temperatures. A 10% drop in temperature results in an ~12% drop in cell performance due to the increase in internal resistance to the flow of oxygen ions. The high temperature also demands that the system include significant thermal shielding to protect personnel and to retain heat. Also, the materials required for such high-temperature operation, particularly for interconnect and construction materials, are very expensive. Operating the SOFC at temperatures lower than 700°C would be very beneficial as low-cost metallic materials, such as ferritic stainless steels, that can be used as interconnect and construction materials. This will make both the stack and balance of a plant cheaper and more robust. Using ferritic materials also significantly reduces the problems associated with chromium. The other advantages of low/intermediate-temperature operation are rapid start-up and shutdown and significantly reduced corrosion rates.

However, to operate at reduced temperatures, several changes are required in stack design, cell materials, reformer design and operation, and operating conditions. With the reduction in operating temperature, the ionic conductivity of the electrolyte decreases and the parasitic losses due to the conductivity of the electrodes and interconnects increase. This results in a rapid deterioration of the performance of the SOFC. This can be overcome to some extent by reducing the thickness of the electrolyte to compensate for its reduced ionic conductivity. The thickness reduction that is required to accommodate, say a 200°C reduction in the operating temperature, leads to impracticably thin membranes.

Some designs in which a thin, dense layer of the electrolyte is physically supported on one of the electrodes (electrode-supported design) are suggested. This structure of a very porous support is difficult to manufacture, and an expensive thin-film deposition technique such as chemical vapor deposition (CVD) is needed to manufacture these systems. Even then, the mechanical strength of the structure (defined by the porous electrode) is often poor, and the handling of the structure through subsequent processing and assembly is difficult. Another approach to improve SOFC performance at low operating temperatures is to use different materials for the electrolyte and the electrode. Several materials options are being investigated [2, 6, 26, 27].

9.3.7 Biofuel cells

A biofuel cell operation is very similar to a conventional fuel cell, except that it uses biocatalysts such as enzymes, or even whole organisms instead of inorganic catalysts like platinum, to catalyze the conversion of chemical energy into electricity. They can use available substrates from renewable sources and convert them into benign by-products with the generation of electricity. As mentioned earlier, in recent years, medical science is increasingly relying on implantable electronic devices for treating a number of conditions. These devices demand a very reliable and maintenance-free (any maintenance that might require surgery) power source. Biofuel cells can provide solutions to most of these problems. A biofuel cell can use fuel that is readily available in the body, for example, glucose in the bloodstream, and it would ideally draw on this power for as long as the patient lives. Since they use concentrated sources of chemical energy, they can be small and light.

A biofuel cell can operate in two ways: It can utilize the chemical pathways of living cells (microbial fuel cells), or, alternatively, it can use isolated enzymes [7, 28]. Microbial fuel cells have high efficiency in terms of conversion of chemical energy into electrical energy; however, they suffer from the low volumetric catalytic activity of the whole organism and low power densities due to slow mass transport of the fuel across the cell wall. Isolated enzymes extracted from biological systems can be used as catalysts to oxidize fuel molecules at the anode and to enhance oxygen reduction at the cathode of the biofuel cell. Isolated enzymes are attractive catalysts for biofuel cells due to their high catalytic activity and selectivity. The theoretical value of the current that can be generated by an enzymatic catalyst with an activity of 10^3 U/mg is 1.6 A, a catalytic rate greater than platinum! However, practical observed currents are much lower due to the loss of catalytic activity from immobilization of the enzymes at the electrode surface and energy losses of the overall system. A major challenge in the biofuel cell design is the electrical coupling of the biological components of the system with the fuel cell electrodes. Molecules known as electron-transfer mediators are needed to provide efficient transport of electrons between the biological components (enzymes or microbial cells) and the electrodes of the biofuel cell. Integrated biocatalytic systems that include biocatalysts, electron-transfer mediators, and electrodes are under research and development. Biofuel cells have much wider fuel options; enzymatic biofuel cells can operate on a wide variety of available fuels such as ethanol, sugars, or even waste materials.

A basic microbial biofuel cell consists of two compartments, an anode compartment and a cathode compartment, separated by a PEM as shown in Fig. 9.11. Usually, Nafion-117 film (an expensive material) is used as the PEM; it allows hydrogen ions generated in the anode compartment to be transferred across the membrane into the cathode compartment [8].

Previously, graphite electrodes were used as the anode and cathode, but they are now replaced by woven graphite felt as it provides a larger surface area than a regular graphite electrode of similar dimensions. This facilitates an increased electron transfer from the microorganisms. A microorganism (e.g., *Escherichia coli*) is used to breakdown glucose in order to generate adenosine triphosphate (ATP), which is utilized by cells for energy storage. Methylene blue (MB) or neutral red (NR) is used as an electron mediator to efficiently facilitate the transfer of electrons from the microorganism to the electrode. Electron mediators tap into the electron transport chain, chemically reducing nicotinamide adenine dinucleotide (NAD$^+$) to its protonated form NADH. The exact mechanism by which the transfer of electrons takes place through these electron mediators is not fully known [29]; however, it is known that they insert themselves into the bacterial membrane and essentially "hijack" the electron transport process of glucose metabolism of the bio-electrodes in a biofuel cell. Their activity is very dependent on pH, and a potassium phosphate buffer (pH 7.0) is used to maintain the pH value in the anode compartment. The cathode compartment contains potassium ferricyanide,

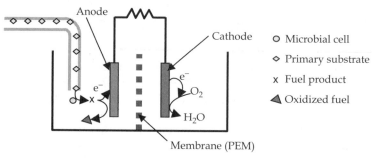

Figure 9.11   Biofuel cell.

a potassium phosphate buffer (pH 7.0), and a woven graphite felt electrode. Potassium ferricyanide reaction helps in rapid electron uptake. Hydrogen ions (H^+) migrate across the PEM and combine with oxygen from air and the electrons to produce water at the cathode. The cathode compartment has to be oxygenated by constant bubbling with air to promote the cathode reactions. It may be worth mentioning that the electron transport chain occurs in the cell membrane of prokaryotes (a unicellular organism having cells lacking membrane-bound nuclei, such as bacteria), while this process occurs in the mitochondrial membrane of eukaryotes (animal cells). Therefore, attempts to substitute eukaryotic cells for bacterial cells in a biofuel cell may present a significant challenge.

Electrochemistry of microbial fuel cells.   In a microbial fuel cell, two redox couples are required in order to generate a current: (a) coupling of the reduction of an electron mediator to a bacterial oxidative metabolism and (b) coupling of the oxidation of the electron mediator to the reduction of the electron acceptor on the cathode surface. The electron acceptor is subsequently regenerated by the presence of O_2 at the cathode surface. The electrochemical reactions in a biofuel cell using glucose as a fuel are

At the anode:

$$C_6H_{12}O_6 + 6H_2O \rightarrow 6CO_2 + 24e^- + 24H^+$$

At the cathode:

$$4Fe(CN)_6^{3-} + 4e^- \rightarrow 4Fe(CN)_6^{4-}$$

$$4Fe(CN)_6^{4-} + 4H^+ + O_2 \rightarrow 4Fe(CN)_6^{3-} + 2H_2O$$

Complete oxidation of glucose does not always occur. One might often get additional products besides CO_2 and water. For example, *E. coli* forms acetate, being unable to completely breakdown glucose, thereby limiting electricity production. Recently, an elegant approach to address this long-standing problem of limited enzyme stability has been reported [30]. It is suggested that the immobilization of enzymes in Nafion layers to create a bio-anode results in stable performance over months.

Another way of using a microorganism's ability to produce electrochemically active substances for energy generation is to combine a bioreactor with a biofuel cell or a hydrogen fuel cell. The fuel can be produced in a bioreactor at one place and transported to a (H_2 or bio-) fuel cell to be used as a fuel. In this case, the biocatalytic microbial reactor produces the fuel, and the biological part of the device is not directly integrated with the electrochemical part (see Fig. 9.12).

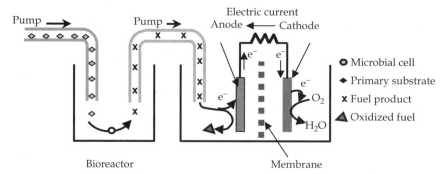

Figure 9.12   Bioreactor and biofuel cell combination.

The advantage of this scheme is that it allows the electrochemical part to operate under conditions that are not compatible with the biological part of the device. The two parts can even be separated in time, operating completely independently. The most widely used fuel in this scheme is hydrogen gas, allowing well-developed and highly efficient H_2/O_2 fuel cells to be conjugated with a bioreactor.

In recent years, ethanol has been developed as an alternative to the traditional methanol-powered biofuel cell due to the widespread availability of ethanol for consumer use, its nontoxicity, and increased selectivity by alcohol. Ethanol fuel cells with immobilized enzymes have provided higher power densities than the latest state-of-the-art methanol biofuel cells. Open-circuit potentials ranging from 0.61 to 0.82 V and power densities of 1.00–2.04 mW/cm^2 have been produced.

Mediatorless microbial fuel cells.   Most biofuel cells need a mediator molecule to speed up the electron transfer from the enzyme to the electrode. Recently, mediatorless microbial fuel cells have been developed. These use metal-reducing bacteria, such as members of the families Geobacteraceae or Shewanellaceae, which exhibit special cytochromes bound to their membranes. These are capable of transferring electrons to the electrodes directly. *Rhodoferax ferrireducens*, an iron-reducing microorganism, has the ability to directly transfer electrons to the surface of electrodes and does not require the addition of toxic electron-shuttling mediator compounds employed in other microbial fuel cells. Also, this metal-reducing bacterium is able to oxidize glucose at 80% electron efficiency (other organisms, such as *Clostridium* strains, oxidize glucose at only 0.04% efficiency). In other fuel cells that use immobilized enzymes, glucose is oxidized to gluconic acid and generates only two electrons, whereas in microbial fuel cells (MFCs) using *R. ferrireducens*, glucose is completely oxidized to CO_2 releasing 24 electrons. These MFCs have a remarkable long-term stability, providing a steady electron flow over extended periods. Current density of 31 mA/m^2 over a period of more than 600 h has been reported [31]. MFCs using *R. ferrireducens* have the ability to be recharged, and have a reasonable cycle life and low capacity loss under open-circuit conditions. They allow the harvest of electricity from many types of organic waste matter or renewable biomass. This is an advantage over other microorganisms in the family Geobacteraceae, which cannot metabolize sugars.

Another recent development has been the use of microfibers rather than flat electrodes and the enzyme-based electroactive coatings. The anode coating used is glucose oxidase, which is covalently bound to a reducing-potential copolymer and has osmium complexes attached to its backbone. The cathode coating contains the enzyme laccase and an oxidizing-potential copolymer. The osmium redox centers in the coatings electrically "wire" the reaction centers of the enzymes to the carbon fibers. This electrode design avoids glucose oxidation at the cathode and O$_2$ reduction at the anode, eliminating the need for an electrode-separating membrane. This has led to miniature "one-compartment bio-fuel cells" for implantable devices within humans, such as pacemakers, insulin pumps, sensors, and prosthetic units. Biofuel cells with two 7-μm-diameter, electrocatalyst-coated carbon fiber electrodes placed in 1-mm grooves machined into a polycarbonate support with a power output of 600 nW at 37°C (enough to power small silicon-based micro-electronics) have been reported [32].

Microbial fuel cells have a long way to go before they compete with more established hydrogen fuel cells or electrical batteries. However, a number of factors provide motivation for research into microbial fuel cells for electricity production.

1. Bacteria are adapted to feeding on virtually all available carbon sources (carbohydrates or more complex organic matter present in sewage, sludge, or even marine sediments). This makes them potential catalysts for electricity generation from organic waste.

2. Bacteria are omnipresent in the environment and are self-reproducing, self-renewing catalysts; thus a simple initial inoculation of a suitable strain could be cultured continuously in an MFC for long-term operation.

3. The catalytic core of conventional fuel cells uses very expensive precious metals such as platinum, and biocatalysts like bacteria may become a serious cost-reducing alternative.

Although biofuel cells are still in an early stage of development and work toward optimizing the performance of a biofuel cell system is needed, the utilization of white blood cells as a source of electrons for a biofuel cell could mark an important step in developing a perpetual power source for implantable devices. There is still a lot of work to be done as there are many unanswered questions; however, the feasibility of constructing commercially viable biofuel cell power supplies for a number of applications is very promising.

9.4   Fuel Cell System

A fuel cell power system requires the integration of many components. The fuel cell produces only dc power and utilizes only certain processed fuels. Besides the fuel cell stack, various components are incorporated in a fuel cell system. A fuel processor is required to allow operation with conventional fuels; a power conditioner is used to tie fuel cells into the ac power grid or distributed generation system; for high-temperature fuel cells, a cogeneration or bottoming cycle plant is needed to utilize rejected heat for achieving high efficiency. A schematic of a fuel cell power system with interaction among various components is shown in Fig. 9.13.

9.4.1   Fuel processor

A fuel processor converts a commercially available fuel (gas, liquid, or solid) to a fuel gas reformate suitable for the fuel cell use. Fuel processing involves the following steps:

1. **Fuel cleaning**—It involves cleaning and removal of harmful species (sulfur, halides, and ammonia) in the fuel. This prevents fuel processor and fuel cell catalyst degradation.

2. **Fuel Conversion**—In this stage, a naturally available fuel (primarily hydrocarbons such as natural gas, petrol, diesel, ethanol, methanol, biofuels [such as produced from biomass, landfill gas, biogas from anaerobic digesters, syngas from gasification of biomass and wastes] etc.) is converted to a hydrogen-rich fuel gas reformat.

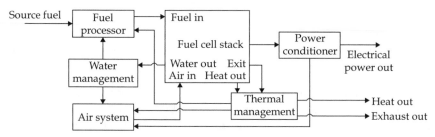

Figure 9.13   A fuel cell power system schematic.

3. **Downstream processing**—It involves reformate gas alteration by converting carbon monoxide (CO) and water (H_2O) in the fuel gas reformate to hydrogen (H_2) and carbon dioxide (CO_2) through the water gas shift reaction, selective oxidation to reduce CO to a few parts per million, or removal of water by condensing to increase the H_2 concentration.

A schematic showing the different stages in the fuel-processing system is presented in Fig. 9.14. Major fuel-processing techniques are steam reforming (SR), partial oxidation (POX) (catalytic and noncatalytic), and autothermal reforming (ATR). Some other techniques such as dry reforming, direct hydrocarbon oxidation, and pyrolysis are also used. Most fuel processors use the chemical and heat energy of the fuel cell effluent to provide heat for fuel processing. This enhances system efficiency.

Steam reforming is a popular method of converting light hydrocarbons to hydrogen. In SR, heated and vaporized fuel is injected with superheated steam (steam-to-carbon molar ratio of about 2.5:1) into a reaction vessel. Excess steam ensures complete reaction as well as inhibits soot formation. Although the steam reformer can operate without a catalyst, most commercial reformers use a nickel- or cobalt-based catalyst to enhance reaction rates at lower temperatures. Although the water gas shift reaction in the steam reformer reactor is exothermic, the combined SR and water gas shift reaction is endothermic. It therefore requires a high-temperature heat source (usually an adjacent high-temperature furnace that burns a small portion of the fuel or the fuel effluent from the fuel cell) to operate the reactor. SR is a slow reaction and requires a large reactor. It is suitable for pipeline gas and light distillates using a fuel cell for stationary power generation but is unsuitable for systems requiring rapid start and/or fast changes in load.

In POX, a substoichiometric amount of air or oxygen is used to partially combust the fuel. POX is highly exothermic, and the resulting high-temperature reaction products are quenched using superheated steam. This promotes the combined water gas shift and steam-reforming

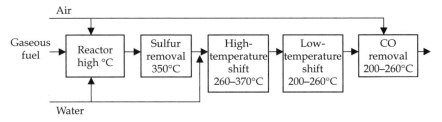

Figure 9.14 A fuel-processing system.

reactions, which cools the gas. In a well-designed POX reformer with controlled preheating of the reactants, the overall reaction is exothermic and self-sustaining. Both catalytic (870–925°C) and noncatalytic (1175–1400°C) POX reformers have been developed for hydrocarbon fuels. The advantage of POX reforming is that it does not need indirect heat transfer, resulting in a compact and lightweight reformer. Also, it is capable of higher reforming efficiencies than steam reformers [3, 6].

Autothermal reforming combines SR with POX reforming in the presence of a catalyst that controls the reaction pathways and thereby determines the relative extents of the POX and SR reactions. The SR reaction absorbs part of the heat generated by the POX reaction, limiting the maximum temperature in the reactor. This results in a slightly exothermic process, which is self-sustaining, and high H_2 concentration. The ATR fuel processor operates at a lower operating cost and lower temperature than the POX reformer, and is smaller, quicker starting, and quicker responding than the SR.

Most of the natural hydrocarbon fuels, such as natural gas and gasoline, contain some amount of sulfur, or sulfur-containing odorants are added to them for leak detection. As the fuel cells or reformer catalysts do not tolerate sulfur, it must be removed. Sulfur removal is usually achieved with the help of zinc oxide sulfur polisher, which removes the mercaptans and disulfides. A zinc oxide reactor is operated at 350–400°C to minimize bed volume. However, removing sulfur-containing odorants such as thiophane requires the addition of a hydrodesulfurizer stage before the zinc oxide polisher. Hydrogen (supplied by recycling a small amount of the natural gas-reformed product) converts thiophane into H_2S in the hydrodesulfurizer. The zinc oxide polisher easily removes H_2S.

To reduce the level of CO in the reformat gas, it must be water gas shifted. The shift conversion is often performed in two or more stages when CO levels are high. A first high-temperature stage allows high reaction rates, while a low-temperature converter allows a higher conversion. Excess steam is used to enhance the CO conversion. In a PEMFC, the reformate is passed through a preferential CO catalytic oxidizer after being shifted in a shift reactor, as a PEMFC can tolerate a CO level of only about 50 ppm.

A fuel processor is an integrated unit consisting of one or more of the above stages, as per the requirements of a particular type of fuel cell. High-temperature fuel cells such as the SOFC and MCFC are equipped with internal fuel reforming and hence do not require a high-temperature shift, or low-temperature shift stage. The CO removal stage is not required for the SOFC, MCFC, PAFC, and circulating AFC. For the PEMFC, all the stages are required.

9.4.2   Air management

Besides fuel, a fuel cell also requires an oxidant (usually air). Depending on the application and design, air provided to the fuel cell cathode can be at a low pressure or a high pressure. High pressure of the air improves the reaction kinetics and increases the power density and efficiency of the stack. But increasing the air pressure reduces the water-holding capacity of the air and therefore reduces the humidification requirements of the membrane (PEMFC). It also increases the power required to compress the air to a high pressure and thereby reduces the net power available. At present, most fuel cell stacks for stationary power applications are designed for operating pressures in the range of 1–8 atm, while automotive fuel cell systems based on the PEMFC technology are designed to operate at lower pressures of 2–3 atm to increase power density and improve water management.

9.4.3   Water management

Water management is critical for fuel cell operation. Water is a product of the fuel cell reaction, and it must be removed from the exhaust gas for use in various operations such as fuel reformation and humidifying reactant gases (to avoid drying out the fuel cell membrane). For automotive applications, water condensed from the exhaust steam is recycled for reforming and reactant humidification in a closed cycle to avoid periodical recharging with water.

9.4.4   Thermal management

The reaction products of the electrochemical reaction in a fuel cell are water, electricity, and heat. The heat energy released in a fuel cell stack is approximately equal to the electrical energy generated and must be managed properly to maintain the fuel cell stack temperature at the optimal level. If this thermal energy (waste heat) is properly utilized, it will considerably increase the efficiency of a fuel cell system. In low-temperature (<200°C) fuel cells (PEMFC, AFC, and PAFC), the stack is cooled by supplying excess air in low power (<200-W) systems, whereas a liquid coolant (deionized water) is used for large-size systems. The waste heat carried out by the coolant is utilized for cogeneration (space heating, water heating, etc.). In high-temperature (<600°C) fuel cell (MCFC and SOFC) systems, all the heat of reaction is transferred to the reactants to maintain the stack temperature at the optimal level. The thermal energy of the high-temperature exhaust may be utilized to preheat the incoming air stream, or in internal or external fuel reformer. The high-temperature exhaust may also be used for cogeneration or electricity generation in a downstream gas turbine system.

9.4.5 Power-conditioning system [33]

The power-conditioning system is an integral part of a fuel cell system. It converts the dc electric power generated by the fuel cell into regulated dc or ac for consumer use. The electrical characteristics of a fuel cell are very far from that of an ideal electric power source. The dc output voltage of a fuel cell stack varies considerably with the load current (see Fig. 9.15), and it has very little overload capacity. It needs considerable auxiliary power for pumps, blowers, and so forth, and requires considerable start-up time due to heating requirements. It is slow to respond to load changes, and its performance degrades considerably with the age of the fuel cell. The various blocks of a fuel cell power-conditioning system are shown in Fig. 9.16.

The dc voltage generated by a fuel cell stack is usually low in magnitude (<50 V for a 5- to 10-kW system, <350 V for a 300-kW system) and varies widely with the load. A dc–dc converter stage is required to regulate and step up the dc voltage to 400–600 V (typical for 120/240-V ac output). Since the dc–dc converter draws power directly from the fuel cell, it should not introduce any negative current into the fuel cell and must be designed to match the fuel cell ripple current specifications. A dc–ac conversion (inverter) stage is needed for converting the dc to ac power at 50 or 60 Hz (see Fig. 9.17). Switching frequency harmonics are filtered out using a filter connected to the output of the inverter to generate a high-quality sinusoidal ac waveform suitable for the load.

9.5 Fuel Cell Applications

The major applications for fuel cells are as stationary electric power plants (including cogeneration units), as a transportation power source

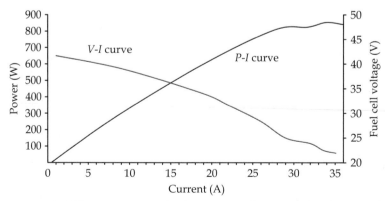

Figure 9.15 Voltage-current and voltage-power characteristics of a typical fuel cell.

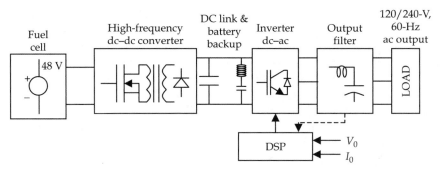

Figure 9.16   Schematic of a fuel cell power-conditioning system.

for vehicles, and as portable power sources, besides an electric power source for space vehicles or other closed environments.

Stationary power applications are very favorable for fuel cell systems. Stationary applications mostly require continuous operation, so start-up time is not a very important constraint. Thus, high-temperature fuel cells such as the MCFC and SOFC systems are also suitable for this application in addition to the PAFC and PEMFC systems. The fuel source for stationary applications is most likely to be natural gas, which is relatively easy to reform in the internal reformer of high-temperature fuel cells or in the external reformer for low-temperature fuel cells. An advantage of using natural gas is that the distribution infrastructure for natural gas already exists. Promising applications for stationary fuel cell systems include premium power systems (high-quality uninterruptible/back-up power supply systems); high-efficiency cogeneration (heat and electricity) systems for residences, commercial buildings, hospitals, and

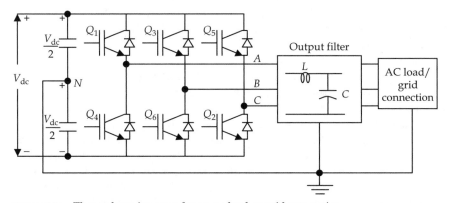

Figure 9.17   Three-phase inverter for an ac load or grid connection.

industrial facilities; and distributed power generation systems for utilities. Although some demonstration and commercial stationary fuel cell power plants in sizes from a few kilowatts to 11 MW are in operation, widespread commercialization can be expected only if their installation cost drops down from the present cost of $4000/kW to about $400–700/kW (or about $1000/kW for some premium applications).

The recent surge of interest in fuel cell technology is because of its potential use in transportation applications, including personal vehicles. This development is being sponsored by various governments in North America, Europe, and Japan, as well as by major automobile manufacturers worldwide, who have invested several billion dollars with the goal of producing a high-efficiency and low-emission fuel cell power plant at a cost that is competitive with the existing internal combustion engines. With hydrogen as the onboard fuel, such vehicles would be zero-emission vehicles. With fuels other than hydrogen, an appropriate fuel processor to convert the fuel to hydrogen will be needed. Fuel cell-powered vehicles offer the advantages of electric drive and low maintenance, because of the few critical moving parts. The major activity in transportation fuel cell development has focused on the polymer electrolyte fuel cell (PEFC), and many of the technical objectives related to the fuel cell stack have been met or are close to being met. The current development efforts are focused on decreasing cost and resolving issues related to fuel supply and system integration.

Besides exotic areas of applications such as space vehicles or submarines, another very promising area of application for fuel cells is portable power systems. Portable power systems are small, lightweight systems that power portable devices (e.g., computers, laptops, cellular phones, and entertainment electronic devices), camping and recreational vehicles, military applications in the field, and so forth. These devices need power in the range of a few watts to a few hundred watts. Fuel cell systems based on DMFC or PEMFC technology are well suited for many of these applications. The convenience of transporting and storing liquid methanol makes DMFC systems very attractive for this application. A small container of methanol or a cylinder of compressed hydrogen can be used as a fuel supply. When the fuel is depleted, a new fuel container may be installed in its place after removing the old one.

In recent years, there has been a lot of interest in electric power generation using renewable energy sources such as wind energy, solar energy, and tidal energy. A major problem with these energy sources is that all are intermittent in nature. Combining the renewable energy-based power generation system with a fuel cell system would solve this problem to a great extent. A hybrid wind/solar energy–fuel cell system can use wind/solar power for generating hydrogen using the electrolysis of water, and store it in cylinders at high pressure. This hydrogen can then be used as the fuel for the fuel cell stack. The stored hydrogen can also be used to fuel the fuel cell vehicles and so forth. In a grid-connected wind/solar energy–hydrogen system, wind/solar power whenever available provides electricity for hydrogen production. The grid power is used during off-peak periods for low-cost electricity and hydrogen production; whereas during peak-demand periods or no/low wind/solar energy periods, the fuel cell can generate electricity using the stored hydrogen. These hybrid systems could be configured in several ways.

9.6 Conclusion

Fuel cell systems are one of the most promising technologies to meet our future power generation requirements. Fuel cell systems provide a very clean and efficient technology for electrical and automotive power systems. With cogeneration efficiencies higher than 80%, fuel cells promise to reduce primary energy use and environmental impact. Fuel cells are a very good alternative for rural energy needs, especially in remote places where there are no existing power grids or power supply is unreliable. The application of fuel cells into the transportation sector will reduce greenhouse emissions considerably; if fuels from renewable energy sources are used, it would nearly eliminate greenhouse gas emissions. Utility companies are beginning to locate small, energy-saving power generators closer to loads to overcome right of way problems and transmission line costs. The modular design of fuel cells suits this distributed generation strategy very nicely as new modular units can be added when the demand increases. This reduces the financial risk for utility planners. Biofuel cells are very attractive for implant devices as they can use glucose in blood to power these devices, eliminating the need for surgery for maintenance and battery replacement. Use of digester gas as a fuel in biofuel cells makes them very attractive for power generation from garbage and other organic waste. This will also help in waste disposal, a big problem in the agriculture and food industry.

All fuel cell technologies (PEMFC, DMFC, AFC, PAFC, MCFC, SOFC, and MFC) discussed in this chapter are in a very advanced stage of development and are very near to commercialization. Although a number of demonstration units of different types of fuel cells are operating all over the world and many PAFC and AFC units have been commercially sold and are successfully operating, fuel cells are still awaiting widespread commercialization due to their high cost and limitation in the choice of the fuel used. These barriers will be overcome in the next few years, and fuel cells will become a preferred power source with widespread applications.

References

1. G. Hoogers (Ed.). *Fuel Cell Technology Handbook*, Boca Raton, Florida: CRC Press, 2003.
2. W. Vielstich, A. Lamn, and H. A. Gasteiger (Eds.). *Handbook of Fuel Cells: Fundamentals, Technology and Applications*, Four Volumes, New York: John Wiley, 2003.
3. Department of Energy. *Fuel Cell Handbook*, sixth ed., Pittsburgh, PA: National Energy Technology Laboratory, DOE, November 2002.
4. M. C. Williams. Fuel cells and the world energy future, *IEEE Power Engineering Society Summer Meeting* **1**, 725, July 15–19, 2001.
5. M. A. Laughton. Fuel cells, *Engineering Science and Education Journal* **11**(1), 7–16, 2002.
6. R. K. Shah. Introduction to fuel cells, In: *Recent Trends in Fuel Science and Technology*, Basu, S. (Ed.), New Delhi, India: Anamya Publishers, pp. 1–9, 2007.
7. R. M. Allen and H. P. Bennetto. Microbial fuel cells—Electricity production from carbohydrates, *Applied Biochemistry and Biotechnology* **39/40**, 27–40, 1993.
8. G. A. Justin. Biofuel Cells as a Possible Power Source for Implantable Electronic Devices, M.S. Thesis, School of Engineering, University of Pittsburg, Germany, 2001.
9. J. Larminie and A. Dicks, *Fuel Cell Systems Explained*, West Sussex, England: John Willey & Sons Ltd., 2003.
10. A. Henzel, R. Nolte, K. Ledjeff-Hey, and M. Zedde. Membrane fuel cells—Concepts and design, *Electrochimica Acta* **43**, 3817–3820, 1998.
11. K. S. Dhathathreyan and N. Rajalakshmi. Polymer electrolyte membrane fuel cell, In: *Recent Trends in Fuel Science and Technology*, Basu, S. (Ed.), New Delhi, India: Anamya Publishers, pp. 41–115, 2007.
12. D. E. Curtin, R. D. Lousenberg, T. J. Henry, P. C. Tangeman, and M. E. Tisack. Advanced materials for improved PEMFC performance and life, *Journal of Power Sources* **131**, 41–48, 2004.
13. S. Litster and G. McLean. PEM fuel cell electrodes, *Journal of Power Sources* **130**, 61–76, 2004.
14. E. Antolini. Recent developments in polymer electrolytic fuel cell electrodes, *Journal of Applied Electrochemistry* **34**, 563–576, 2004.
15. M. Daugherty, D. Haberman, N. Stetson, S. Ibrahim, D. Lokkon, D. Dunn et al. Modular PEM fuel cell for outdoor applications, In: *Proceedings of the European Fuel Cell Forum Portable Fuel Cells Conference*, 1999, Lucerne, Switzerland, pp. 205–213.
16. I. Bar-On, R. Kirchain, and R. Roth. Technical cost analysis of PEM fuel cells, *Journal of Power Sources* **109**, 71–75, 2002.
17. S. Venugopalan. Micro fuel cells, In: *Recent Trends in Fuel Science and Technology*, Basu, S. (Ed.), New Delhi, India: Anamya Publishers, pp. 137–156, 2007.
18. T. Burchardt, P. Gouérec, E. Sanchez-Cortezon, Z. Karichev, and J. H. Miners. Alkaline fuel cells: Contemporary advancement and limitations, *Fuel* **81**, 2151–2155, 2002.
19. G. F. McLean, T. Niet, S. Prince-Richard, and N. Djilali. An assessment of alkaline fuel cell technology, *International Journal of Hydrogen Energy* **27**, 507–526, 2002.
20. E. Gülzow and M. Schulze. Long-term operation of AFC electrodes with CO_2 containing gases, *Journal of Power Sources* **127**, 243–251, 2004.
21. M. Cifrain and K. V. Kordesch. Advances, aging mechanism and lifetime in AFCs with circulating electrolytes, *Journal of Power Sources* **127**, 234–242, 2003.
22. S. Roy Choudhury. Phosphoric acid fuel cell technology, In: *Recent Trends in Fuel Science and Technology*, Basu, S. (Ed.), New Delhi, India: Anamya Publishers, pp. 188–216, 2007.
23. W. Vielstich, A. Lamm, and H. A., Gasteiger (Eds.). *Handbook of Fuel Cells, Fundamentals, Technology and Applications*, Vol. 1, West Sussex, England: John Wiley and Sons Ltd., 2003.
24. H. Ghezel-Ayagh, M. Farooque, and H. C. Maru. Carbonate fuel cell: Principles and applications, In: *Recent Trends in Fuel Science and Technology*, Basu, S. (Ed.), New Delhi, India: Anamya Publishers, pp. 217–247, 2007.

25. M. Farooque, S. Katikaneni, and H. C. Maru. The direct carbonate fuel cell technology and products review, In: *Carbonate Fuel Cell Technology V, Electrochemical Society Proceedings,* 1999, **99–20**, pp. 47–65.

26. R. Bove. Solid oxide fuel cells: Principles, design and state-of-the-art in industries, In: *Recent Trends in Fuel Science and Technology*, Basu, S. (Ed.), New Delhi, India: Anamya Publishers, pp. 267–285, 2007.

27. A. J. Appelby and F. R. Foulkes. *Fuel Cell Handbook*, New York: Van Nostrand, 1989.

28. G. T. R. Palmore and G. M. Whitesides. Microbial and enzymatic biofuel cells, In *Enzymatic Conversion of Biomass for Fuels Production*, Himmel, M. E., Baker, J.O., and Overend, R. P. (Eds.), ACS Symposium Series No. 566, Washington, DC: American Chemical Society, pp. 271–290, 1994.

29. B. E. Logan, B. Hamelers, R. Rozendal, U. Schröder, J. K. S. Freguia, P. Aelterman, W. Verstraete, and K. Rabaey. Microbial fuel cells: Methodology and technology, *Environmental Science and Technology* **40**(17), 5181–5192, 2006.

30. K. Rabaey, G. Lissens, S. D. Siciliano, and W. Verstraete. A microbial fuel cell capable of converting glucose to electricity at high rate and efficiency, *Biotechnology Letters* **25**, 1531–1535, July 2003.

31. S. K. Chaudhuri and D. R. Lovley. Electricity generation by direct oxidation of glucose in mediatorless microbial fuel cells, *Nature Biotechnology* **21**(10), 1229–1232, 2003.

32. T. Chen, S. C. Barton, G. Binyamin, Z. Gao, Y. Zhang, H.-H. Kim, and A. Heller. A miniature biofuel cell, *Journal of American Oil Chemists' Society* **123**(35), 8630–8631, 2001.

33. S. K. Mazumder. Fuel cell power-conditioning systems, In: *Recent Trends in Fuel Science and Technology*, Basu S. (Ed.), New Delhi, India: Anamya Publishers, pp. 332–355, 2007.

Relevant Definition of Energy/Work Units

Btu British thermal unit. Heat energy necessary to raise the temperature of 1 lb of water 1°F.

cal or gcal Calorie or gram calorie. Heat energy required to raise the temperature of 1 mL of water 1°C (from 15 to 16°C).

electron volt 1.6×10^{-12} erg $= 1.6 \times 10^{-19}$ J $= 23.06$ kcal/mol. Energy gained by an electron passing through a potential of 1 V.

foot · pound ft · lb. Work energy needed to raise 1 lb to a height of 1 ft $= 0.138$ kg · m.

force Correct force definition can be obtained from the second law of Newton stating that the inertia is disturbed by unbalanced force, which causes acceleration on a body directly proportional to the force (F) and inversely proportional to the mass of the body $F = K\,mf$ ($F = Kmf$ where m is mass and f is acceleration). If all are reduced to unity, the unit of force becomes pound foot per second per second (poundal) or gram centimeter per second per second (dyne).

joule Work energy to raise 1 kg to a height of 10 cm $= 0.1$ kg · m $= 0.74$ ft · lb.

joule (electrical) 0.239 cal. Energy developed when 1 C of electrons (10.364×10^{-6} mole) passes through a potential of 1 V.

kcal/einstein Energy of a mole of a photon (einstein) of wavelength (in μ) $28589.7\ \mu^{-1}$ kcal/mol.

power Rate of doing work, $P = W/t$.

$$J/s = W \text{ or ft} \cdot lb/s$$

$$\text{or horsepower} = 550 \text{ ft} \cdot lb/s \text{ or } 33{,}000 \text{ ft} \cdot lb/min.$$

So, $1 \text{ W} = 10^7 \text{ ergs/s} = 1 \text{ J/s} = 0.239 \text{ cal/s}$

$1 \text{ hp} = 550 \text{ ft} \cdot \text{lb/s} = 33{,}000 \text{ ft} \cdot \text{lb/min}$

$1 \text{ hp} = 746 \text{ W} = 178 \text{ cal/s}$

$1 \text{ kW} = 1000 \text{ W} = 1.34 \text{ hp}$

$1 \text{ kWh} = 3.6 \times 10^6 \text{ J} = 860 \text{ kcal/h} = 3413 \text{ Btu/h}$

$1 \text{ ft} \cdot \text{lb/s} = 1.356 \text{ W} = 0.324 \text{ cal}$

quantum   Wavelength $\mu\text{eV} = 1239.8 \ \mu^{-1}$

Wave number $\text{cm}^{-1} \text{eV} = 1239.8 \times 10^{-4} \text{ cm}^{-1}$

units $1 \text{ erg} = 1 \text{ dyne} \times 1 \text{ cm} \ (\text{dyne} = 1 \text{ g} \cdot \text{cm s}^{-2})$

$1 \text{ J} = 10^7 \text{ ergs} = 0.74 \text{ ft} \cdot \text{lb} = 0.239 \text{ cal}$

$1 \text{ ft} \cdot \text{lb} = 1.3549 \text{ J}$

work, W   Force $\times$ distance (both in the same direction on a body).

Index